A Challenge to the Whole Physics World

A Challenge to the Whole Physics World

K.H.K. Geerasee Wijesuriya

**BSc (Hons) in Physics and Mathematics,
Faculty of Science, University of Colombo**

A Challenge to the Whole Physics World
© 2016 K.H.K. Geerasee Wijesuriya. All rights reserved

ISBN-13: 978-1533012944
ISBN-10: 1533012946

1st Edition: May 2016

Library of Congress Control Number (LCCN): 2016907545
CreateSpace Independent Publishing Platform, North Charleston, SC,

www.Amazon.com, CreateSpace United States of America.

Cover photo origin credit: https://en.wikipedia.org/wiki/Black_hole

Author's Biography

The author of this research article is K.H.K. Geerasee Wijesuriya (Normally identify as Geerasee Wijesuriya). And the ideas in this book are innovative and those are completely K.H.K. Geerasee Wijesuriya's arguments. Geerasee studied at Faculty of Science, University of Colombo Sri Lanka. And she graduated with BSc(Hons) in Physics and Mathematics from the University of Colombo, Sri Lanka in 2014 June.

Geerasee has been invited by several Astronomy/Physics institutions and organizations world-wide, asking to get involve with them. Also, She has received several invitations from some private researchers around the world asking to contribute to their researches.

She worked as Mathematics tutor/Instructor at Mathematics department, Faculty of Engineering, University of Moratuwa, Sri Lanka. Also now she is intending to complete her PHD in Physics and Astronomy.

Acknowledgement

I would be thankful to my parents who gave me the strength to move forward with physics knowledge and achieve my scientific goals.

List of abbreviations

Faculty of Science, University of Colombo, Sri Lanka.

Preface

The present human generation is the current form of the long chain of evolution of human being. With the evolution of the mankind, human's brain power took the major place among other species. Due to the evolution of the brain power of human kind, extraordinary observations came to the world. People consequently observed some special features around them in the surrounding environment of the universe. And they thought how to use those unusual observations to make their life easy. Gradually people tried to think differently rather than other animal species. That was the main difference between human kind and the other animals.

In the ancient era, people tried to make their life easier through several procedures. They figured out how to make levers, fire and some cooking equipment. They used those fundamental scientific methods to advance their life.

With the further evolution of human brain, some people became more genius than others due to some gene differences of their parents. That was the main starting point of the present science subject aspects.

Due to the intellectual differences of humans, educational and social power variances came to the human world. Then gradually educational institutions and new scientific equipment came to the picture across the world. People who tried to make their life more advance and easy recognized as scientists and they took the more important position in the human strategy.

During the 14-19 century, some important scientific discoveries came to the science world. Among those scientists who engaged with those discoveries, Sir Isaac Newton, Galileo Galilei, Johannes Kepler were some of them. The 18th - 19th centuries usually named as the mechanical era of the human evolution.

But in the 20th century, another most challengable scientist took the position in the Physics world. His name as we all know is Albert Einstein.

Einstein's most important talent was, he was a man who capable to make mind-based experiments without any physical scientific equipment or any laboratory. His Special theory of relativity and the General theory of relativity brought a revolution to the whole physics and other science worlds. According to several science authors, Einstein should be the most attractive scientist that they have ever met in the science history of the world.

Einstein's theories are very important to build several mechanical equipment and specially, for nuclear bombs. Einstein was a scientist who was capable to identify the connection between the space-time and the matter-energy. Since majority of Einstein's theories are theoretical, there were no any easy experimental procedures to test the validity of those theories. But Einstein found a way to test his innovative theories experimentally under some special circumstances. Anyway finally his arguments made the reality to the Physics World.

Einstein made the Newton's concepts limited and he proved that Newton made the reality only under classical mechanical circumstances. But recent scientists expressed a challenge to Einstein's arguments through several experiments and theories.

The main attempt of this book is to point out some special contradictory facts in Einstein's concepts those have the capability to carry revolution to the physics world again.

This book contains challengable concepts to the current physics as well as new theoretical procedures those have directly related with Astrophysics. This book is so worthy for Physics Undergraduates, Graduates and scientists and as well as any other researcher who has interest in innovative ideas.

K.H.K. Geerasee Wijesuriya

Author's email address: geeraseew@gmail.com
Author's postal address: 86, Sucharitha Mawatha, Nawinna, Maharagama, SriLanka

Content

A challenge to the whole physics world

Chapter 01

An indirect evidence for the existence of Dark energy/Dark matter by considering a pulsar

1.1.   What is a pulsar

We all know that a pulsar is a rapidly rotating neutron star. Especially, the density of a pulsar is very high. Because, it consists of almost neutrons with high concentration densities.

Similar to Earth, pulsar has axis of rotation and magnetic axis. But pulsar emits high energy electromagnetic waves along the direction of the magnetic axis. Usually those EM waves are highly energized Gamma radiations. Since the pulsar emits highly energized gamma radiations along the magnetic axis, an observer able to see the pulsar's photons discretely. Because the axis of rotation and magnetic axis are different in the alignments. Then the magnetic axis creates a symmetric cone which is symmetric along the axis of rotation. Then only when the magnetic axis is towards the Earth, observer on Earth can see the photons coming from the pulsar.

But not otherwise.

1.2.   How to get an indirect evidence of a pulsar

My attempt is to investigate my own ideas those related with several cosmological subject fields.

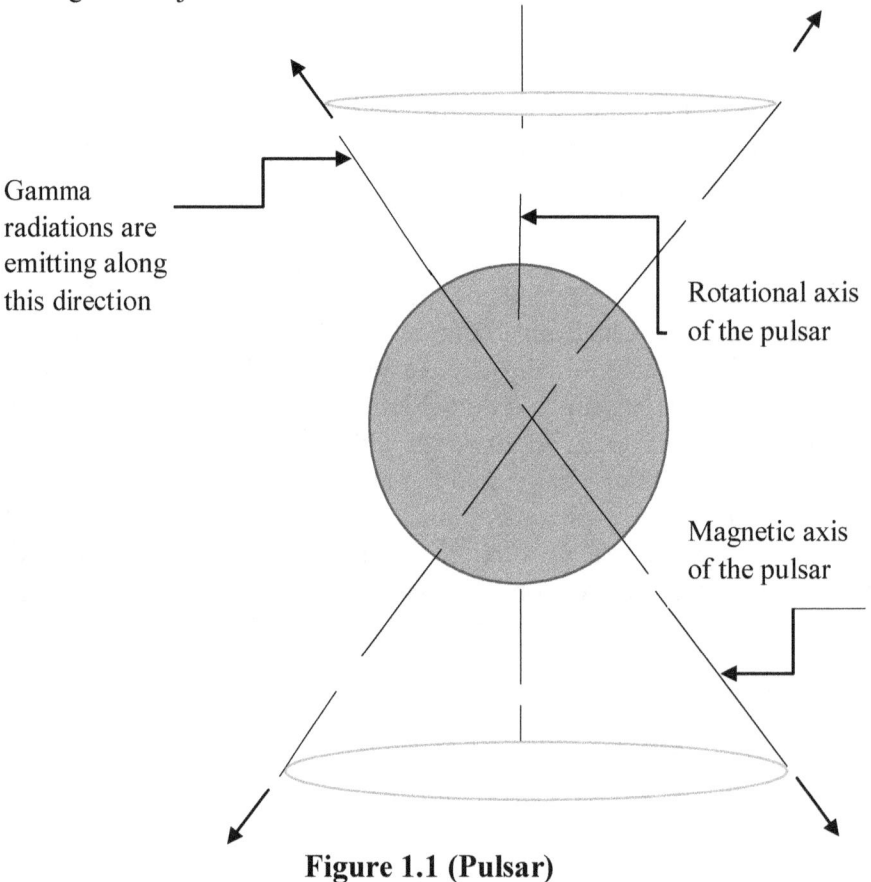

Gamma radiations are emitting along this direction

Rotational axis of the pulsar

Magnetic axis of the pulsar

Figure 1.1 (Pulsar)

Let's consider the time duration between two consecutive detections of gamma photons is $\Delta t > 0$.

Then,

The angular velocity of the pulsar = $\omega = 2*\pi/\Delta t$ 1.1

Photo credit: NASA/Fermi/Cruz de
Figure 1.2: Pulsar and its Gamma radiations emission

Usually, the pulsars those we can observe are very distant objects from the Earth. Then there is a relative velocity due to the universal expansion according to the Big Bang theory. That means the universal expansion causes a red shift of the incoming photons from the pulsar to the Earth. But due to the tangential velocity component at the magnetic pole of the pulsar when the gamma radiations were emitting, there is another red shift/ blue shift of the incoming gamma photons.

You all know that scientists able to use a spectrometer to analyze the electromagnetic waves coming from the distant universe. By using the resolution images, they are capable to identify the chemicals in the distant object's surface if they analyze the emission spectrum. Similar to that, after analyzing the emission spectrum of the pulsar, researcher is capable to calculate the wavelengths of two different regions of the emission spectrum. But according to the Doppler Effect, researcher cannot estimate the exact wavelengths of the chemicals of the pulsar by neglecting the Doppler Effect. Thus he should take red shift due to the universal expansion plus the red shift/blue shift due to the pulsars rotation along the axis of rotation.

Let's consider an emission spectrum of the pulsar's EM waves that has taken at time T.

Let, f_1 is the frequency of emission line L_1 that has estimated using the emission spectrum taken at time T. f_0 is the exact frequency of the considering photon that has emitted by the pulsar.

Let, f_2 is the frequency of emission line L_2 that has estimated using the emission spectrum taken at time T. f'_0 is the exact frequency of the considering other photon that has emitted by the pulsar.

But the photons associated with line L_1 and L_2 have received simultaneously. Then, according to the relativistic Doppler Effect:

$$f_1 = f_0 * \sqrt{\frac{1 - \frac{Vb + Vp.\cos\theta}{C}}{1 + \frac{Vb + Vp.\cos\theta}{C}}} \dots\dots\dots\dots\dots\dots\dots\dots\dots 1.2$$

Where, V_b is the relative velocity of the pulsar due to the universal expansion that might caused by the Big Bang incident. Also, V_p is the tangential velocity of the pulsar when the detecting photon was emitting by the pulsar. θ is the angle between the line joining the Earth and the pulsar and the line parallel to the tangential velocity direction of the pulsar when the detecting photon was emitting. C is the speed of light.

Also,

$$f_2 = f'_0 * \sqrt{\frac{1 - \frac{Vb + Vp.\cos\theta}{C}}{1 + \frac{Vb + Vp.\cos\theta}{C}}} \dots\dots\dots\dots\dots\dots\dots\dots 1.3$$

Depending on the detection of the photon at time T, V_p may negative or positive.

By equation (2) and (3): $f_1/f_2 = f_0/f'_0 \dots\dots\dots\dots\dots\dots\dots\dots 1.4$

By analyzing the spectrum, we can calculate the value of f_1/f_2. Let's consider $f_1/f_2 = x$.

Then $f_0/f'_0 = x$. But f_0 and f'_0 are non-shifted frequency values. Then by comparing laboratory measurements of photon frequencies of various chemicals with the value x; we can identify what should be the chemicals associated with the frequency f_0 and f'_0. Then we can estimate the values of f_0 and f'_0. Then by the equation (2), we can calculate the value of

$$\sqrt{\frac{1 - \frac{Vb + Vp.\cos\theta}{C}}{1 + \frac{Vb + Vp.\cos\theta}{C}}} = y \dots\dots\dots\dots\dots\dots\dots\dots 1.5$$

But in order to calculate V_b, we should know the distance between the Earth and the pulsar when the detecting photon (came from the pulsar) was emitting by the pulsar. There are several ways to calculate the distance to a faraway object.

Let's consider the below method:
By using spectral classes, researcher has the capability to identify the color of the particular pulsar and then she able to measure the surface temperature of the particular pulsar. By Hertzsprung-Russell diagram the researcher able to calculate the absolute brightness (M) of the pulsar. After measuring the apparent brightness (m) of the pulsar, by using the equation ($5.\log_{10} D = m-M -10$) he able to calculate the distance to the pulsar (D) at the detecting photon emitted moment.
Especially this method is valid exactly only for the pulsars those are comparably closer to the Earth. Then we can say the value of V_b when the detected photon was emitting by the pulsar and the present moment are same (according to the Hubble's law). Also then only we can say the value of D when the detected photon emitted by the pulsar and the value of D at the present moment are roughly same.

Then by the Hubble's equation for universal expansion, $V_b = H. D$, we can calculate the value V_b

Then by equation (5), we can calculate the value V_p. $\cos \theta = k$...........1.6

But by the diagram number (1); we can identify that we are capable to detect the EM waves coming from the pulsar: only when the tangential velocity direction of the pulsar and the gamma radiations coming direction are perpendicular to each other.

The above statement is true ONLY IF THE CONE (MADE DUE TO THE MOTION OF MAGNETIC AXIS OF THE PULSAR) IS SYMMECTRIC AROUND THE AXIS OF ROTATION OF THE PULSAR.

1.3.   Conclusion of Chapter 01

If $y \neq \sqrt{\frac{1-\frac{Vb}{c}}{1+\frac{Vb}{c}}}$; we can conclude that there is some external force acting on the pulsar or pulsar is influenced by its own change of its internal magnetic force acting on it. By considering the nearby visible matter to the pulsar, the variations of the value 'y' from the value $\sqrt{\frac{1-\frac{Vb}{c}}{1+\frac{Vb}{c}}}$ at different times, spectroscopic data of infrared radiations around the pulsar and the internal magnetic field strength calculated through the variations of the gamma rays coming from the pulsar; researcher is capable to identify whether is there any un-visible matter/energy around the pulsar or not. Because by considering the difference between y and $\sqrt{\frac{1-\frac{Vb}{c}}{1+\frac{Vb}{c}}}$ time to time;

he may able to estimate theoretically the net force(F_1) that should be acting on the pulsar due to the surrounding visible matters. But experimentally the researcher able to calculate the net gravitational force (F_2) acting on the pulsar by observing the surrounding visible matters. If there is any difference between F_1 and F_2; he is capable to conclude that there is some another un-visible gravitational force acting on it. That **should be an indirect evidence for the dark matter and dark energy.** This argument is perfect if he can verify that there is no any change of the internal magnetic force or there is no any other external magnetic force acting on it as below. But if there is any internal (self-exist) change of magnetic field or any external change of magnetic field that is acting on the pulsar, the researcher able to use the Maxwell's laws to detect whether is there any such a change of magnetic field or not. If there is any external/internal magnetic field change of the pulsar, then by analyzing spectrum of the infrared radiations coming from the pulsar's surface; he may calculate the force due to the change of the magnetic field. Then, if the theoretically calculated force associated with the difference

$$| y - \sqrt{\frac{1-\frac{Vb}{c}}{1+\frac{Vb}{c}}} | \neq [\text{ Calculated external gravitational force acting on the}$$

pulsar + calculated force acting due to the change of the magnetic force on the pulsar]
The researcher is proud to investigate that 'The researcher has found indirect evidence for the existence of dark matter/dark energy '.

Moreover, by using luminosity, temperature and other properties of the pulsar, the researcher able to calculate the radius of it. By using equation (1), he may able to calculate the tangential velocity (V_p) of the pulsar when the beam of gamma rays is towards us. Then by using equation (5), he may able to calculate the value of θ. Then that angle value will help the researcher to identify where should be the Dark energy/Dark matter strong in the strength.

Chapter 02

Newton's First Law of motion is not real

Photo origin: www.learnwithmac.com

2.1.Brief recent History of Science

The development of science happens through two major aspects. Those are theoretical and the experimental. Once a scientist investigated some specific phenomenon, other experimental scientists have to confirm that through an experiment. Before hundreds years ago Sir Isaac Newton investigated several groundbreaking scientific ideas (those were suited with that era) regarding the nature of the Universe. Among those innovative ideas, the Newton's Universal theory of gravity, Newton's first law, 2nd law and 3rd law are the concepts those carried a physical revolution to the physics world. With the Newton's Universal theory of gravity, Newton pointed out an idea that was containing the fundamental understanding of the nature of the gravity at that time.

But, in 1916 Albert Einstein re- formulated the understanding of the nature of the gravity by using several mathematical equations and solutions for them. The very important feature of Einstein's investigation was his new idea gathered the understanding of the space-time and the mass-energy. Before Einstein, nobody had gathered the nature of the space-time and the nature of the mass-energy into one single framework. But recent scientists have investigated several new specific ideas those contradict the Einstein's fundamentals through the experiments and through the day-to-day observations as well. With the development of science, scientists have the capability to investigate several new notions (easier than previous) those can break all the basis of the physics subject. They can use the experimental procedures and the theoretical arguments to change the world.

2.2. What is the procedure to contradict Law of Inertia

Similar to Einstein's works, my attempt is to emphasize an incorrectness of Newton's first law of motion. Through this new article my attempt is to give an innovative idea that is carrying a contradiction to the Newton's first law of motion. That is: "If there is not any external force acting on an object, the object should move with a constant velocity or that object should be at rest". But my attempt is to change this idea through a mind-based experiment. I may express this innovative idea by considering an object which is in the empty space-time.

Research Content

Let's consider an object, which has carried to the empty space. Here, empty space is a space-time which does not contain any matter particles. Also, this empty space has not been affected by any external force such as gravitational force, Electromagnetic force, Weak force or Strong nuclear force. After carrying the object to the empty space, we apply some non-zero pushing force to that object at time T=0.

Let's use a diagram to understand the situation.

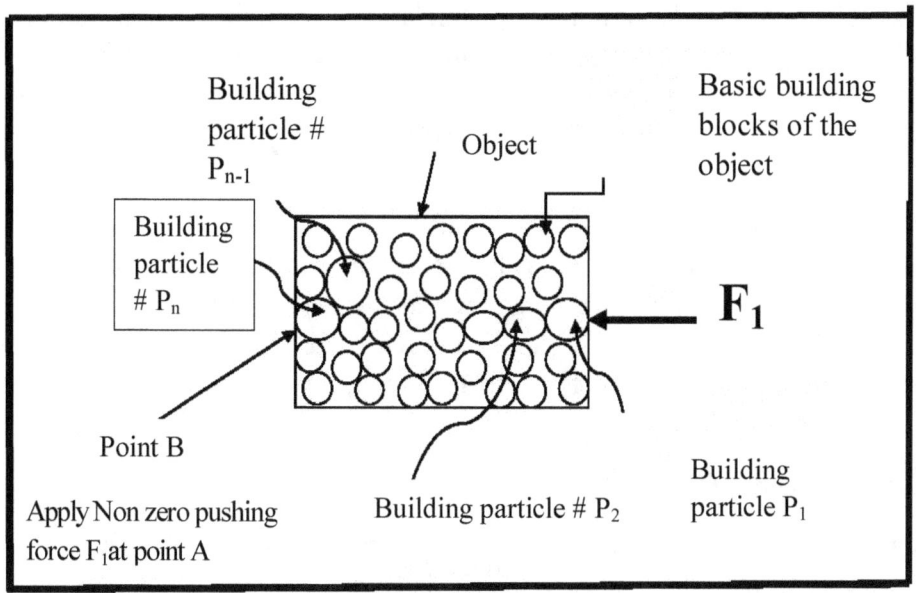

Figure 2.1

The object consists large number of fundamental building matter particles (Those have denoted by P_1, P_2, P_{n-1} and P_n in the figure number 2.1). The dimensions of each fundamental building matter particle cannot be changed under any influences of any external force.

When we apply some non-zero pushing force (F_1) to the object at point A (At point A, the matter particle is P_1), P_1 absorbs some amount energy due to that F_1. We apply the pushing force to the object at time T=0. Again, after Δt time we remove the applied force. After that, we do not again apply any pushing or any force to the object. Then start to count time flowing starting from

$T = T_0$. Where $T(=0) + \Delta t = T_0$.

Let's consider P_1 absorbed E_1 energy due to the applied pushing force (But remember P_1 particle should be the particle which is the nearest particle to the point A). Then P_1 has kinetic energy and then it moves along the direction which was the force applied and passes some amount of energy to the next nearest particle. Let's consider that particle as P_2. Then P_2 also absorbs some amount of energy E_2 (E_2 should be less than E_1) and moves some distance. Again, the particle nearest to P_2 absorbs some amount of energy (E_3) from P_2 and moves. Then, the same chain of procedure happens for all other particles inside the object. **After the most faraway particle (P_n) absorbed some amount of kinetic energy, the whole object start to move through space-time at time T*.** But we cannot observe any time difference between T_0 and the time T* (Time interval that spends to pass the absorbed energy by the particles near point A to the particles near point B) practically. But after the very first process of passing energy; **P_1 has E_1' ($< E_1$) energy** and again **passes some amount** of energy from E_2 to other particles (In the second cycle of passing energy throughout the block of matter) similar to the previous procedure. And, happens the same procedure for all other particles inside the object. But in the very-first cycle of absorbing energy by the particles, P_1 has the highest kinetic energy rather than other particles. Just after P_1 absorbed kinetic energy from the pushing force there may be several particles those **haven't absorbed** any amount of energy due to the pushing force. **But P_1 has**.

Since all the particles have the same mass, P_1 should move faster than other particles (Because $E_1 > E_2 > \ldots\ldots > E_n$; Here E_n is the kinetic energy absorbed by the particle which is most far away from the point A during the very first cycle of energy passing procedure). But all the particles have created a single block of matter which is the object. But during the **very first cycle** of passing the kinetic energy throughout the block of matter, due to the energy differences of the particles, the object should **compress** (Because the particles near the point A has higher energy than the particles near the point B- Then particles near the point A move long distances, but some particles near B haven't absorbed any amount of energy yet OR those particles near B have low amount of kinetic energy).

But during the 2nd time of passing energy through the block of particles (in addition to the newly absorbing energy in the 2nd cycle of passing energy), there are non-zero previous energies (Due to the 1st cycle of passing energy throughout the block of matter) remaining yet with each fundamental particle. i.e. after the first cycle of passing energy among each other by particles, we cannot guarantee any order of energy that they should have (Such as Energy of P1 > Energy of P2 >.......). But due to the concentration system of particles inside the object, every particle move with several velocities depending on the energy they have absorbed through the other particles (Due to the initially absorbed Energy from the pushing force that we applied at time T=0). i.e. depending on the location of each particle inside the object, the size and the volume of the whole object should change. Let's figure out the situation as below.

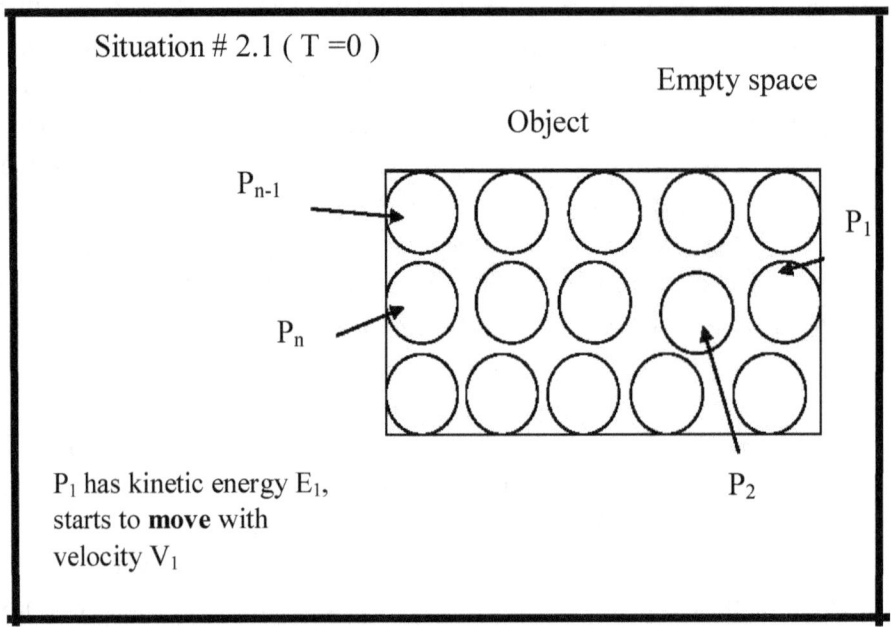

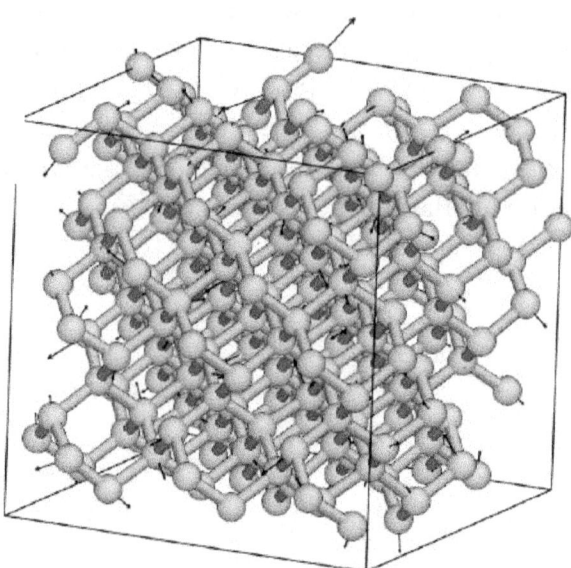

Fundamental particle structure inside an object(Photoorigin: libatoms.github.io)

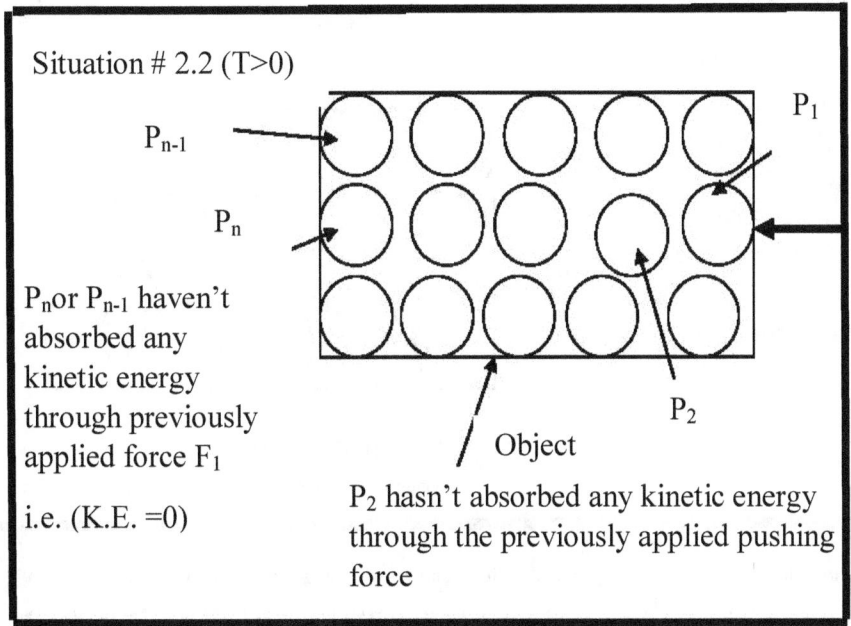

Situation # 2.2 (T>0)

P_{n-1}

P_1

P_n

P_n or P_{n-1} haven't absorbed any kinetic energy through previously applied force F_1

i.e. (K.E. =0)

P_2

Object

P_2 hasn't absorbed any kinetic energy through the previously applied pushing force

Situation number 2.2: There is no any external pushing force

(P_1 starts to move with velocity V_1). But P_2, P_3,......,P_{n-1}, P_n are at rest.

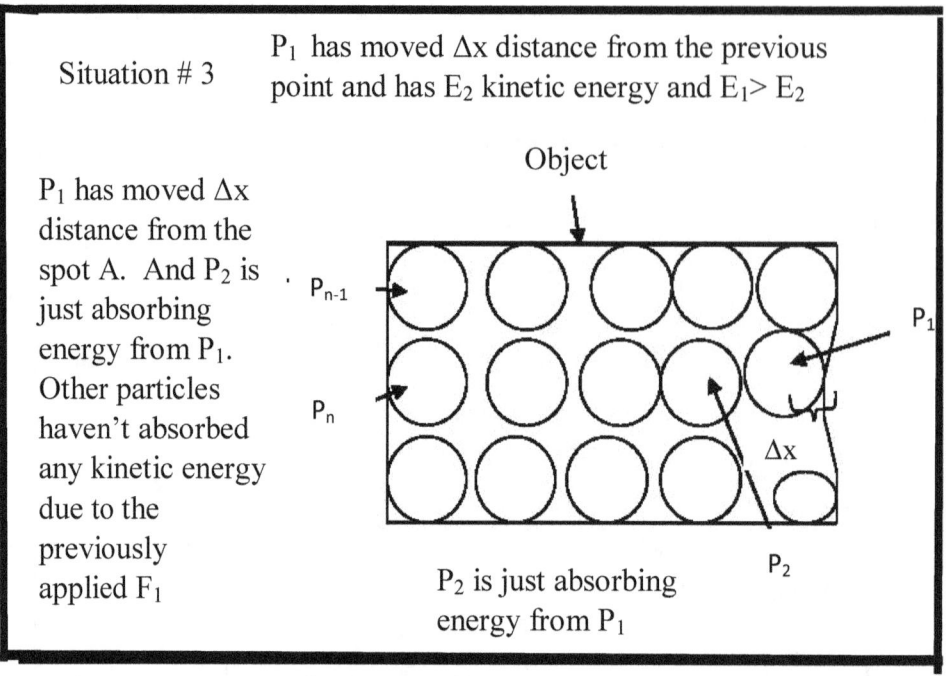

| Situation # 3 | P_1 has moved Δx distance from the previous point and has E_2 kinetic energy and $E_1 > E_2$ |

Object

P_1 has moved Δx distance from the spot A. And P_2 is just absorbing energy from P_1. Other particles haven't absorbed any kinetic energy due to the previously applied F_1

P_2 is just absorbing energy from P_1

Situation number 2.3: The object has further compressed along the direction of previously applied F_1. P_1 is moving with velocity V_1. All other particles are at rest. The shape of the object at point A was determining by P_1. Therefore with the motion of P_1, the shape of the object at point A changes.

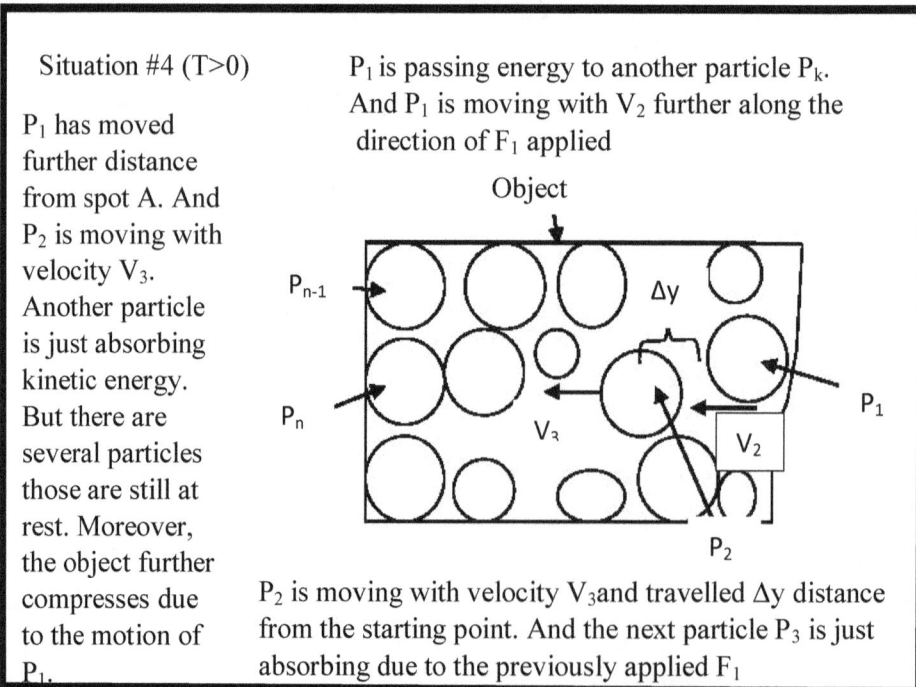

| Situation #4 (T>0) | P_1 is passing energy to another particle P_k. And P_1 is moving with V_2 further along the direction of F_1 applied |

P_1 has moved further distance from spot A. And P_2 is moving with velocity V_3. Another particle is just absorbing kinetic energy. But there are several particles those are still at rest. Moreover, the object further compresses due to the motion of P_1.

Object

P_2 is moving with velocity V_3 and travelled Δy distance from the starting point. And the next particle P_3 is just absorbing due to the previously applied F_1

Therefore, as above described, all the particles inside the object do not start the motions simultaneously after applying some pushing force.

But after all the single building particles of the object absorbed kinetic energy due to F_1, we can't guarantee the order of the kinetic energies each particle has (i.e. Kinetic energy of P_1> Kinetic energy of P_2>.........P_n likewise). But the amount of kinetic energies the particles have at some particular moment after T^* depend on the particle concentration system inside the object. But just before T^* time, P_1, P_2, P_3,....., P_{n-1} all are moving toward the object's moving direction without any negative component.

But after all the fundamental building particles of the object absorbed the kinetic energy from F_1; we can observe that the object is moving along the direction of motion.

But before T^*, there was not any motion of the whole object; but there were several motions near the point A. Therefore, before T^*, the object should compress along the direction of motion. Let's figure out the situation as below.

Figure 2.4

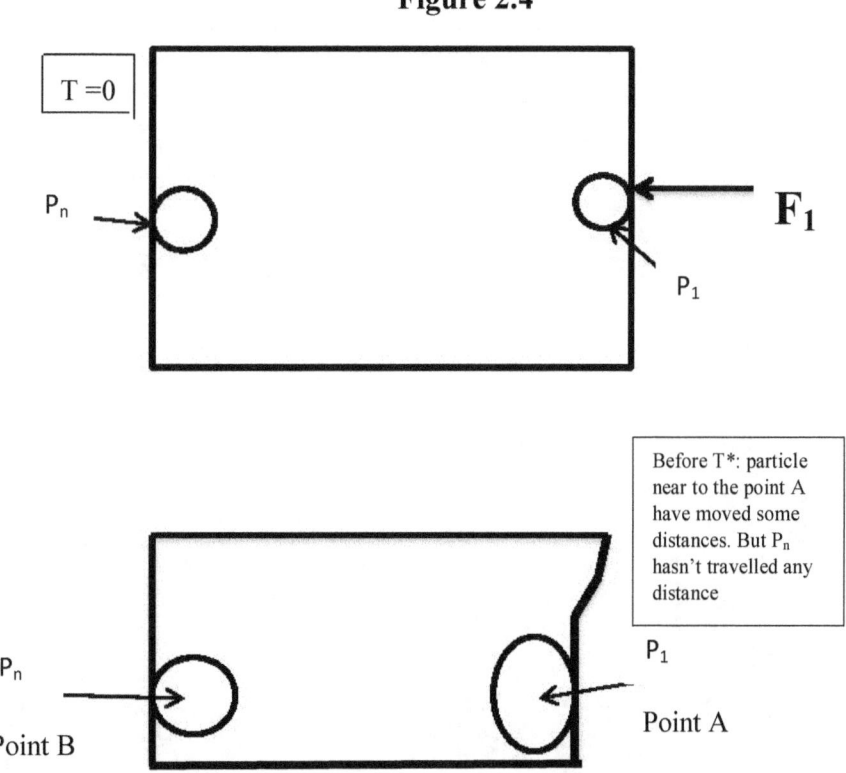

Before T^*: particle near to the point A have moved some distances. But P_n hasn't travelled any distance

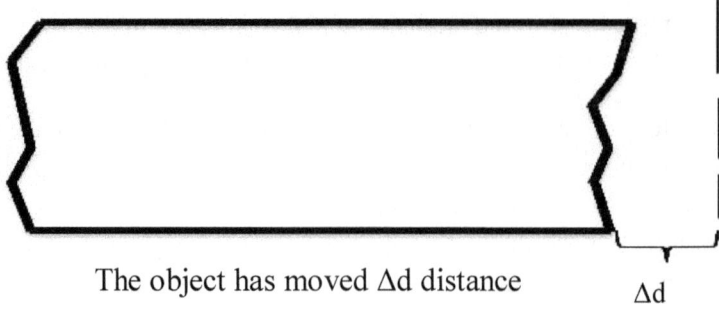

The object has moved Δd distance

Δd

After T*: particles have different amounts of kinetic energies. If particles near to the point B have higher amounts of kinetic energies, than the particles near to the point A, the object should expand along the direction of whole object's motion

Figure 2.5

Distance travelled by an object we measure as below.

We consider some point on the object. Then, we measure the distance between two locations of that particular point for two moments.

But if we consider some point q on the object, that point does not move uniformly throughout unit time intervals after T=0 (That is due to the non-uniform Kinetic Energy amounts of each particle during each unit time interval).

But someone can argue that the gravitational center of the whole object may move uniformly throughout the unit time intervals. But NO. Why? The object sometimes expands. Sometimes compresses (Due to the non-uniform variations of Kinetic energy of the particles near point A and Point B. Then the particle concentration of the object does not vary uniformly throughout the unit time intervals). Then, the gravitational center of the object does not move uniformly throughout unit time intervals. And there is no any point in the object that moves uniformly throughout the unit time intervals (Due to the uncertain expansion and the contraction of the object).

Moreover, someone can say when P_1 absorbed kinetic energy from F_1, it moves some distance towards P_2. Then, due to the repulsive forces presence due to other particles near to P_1, the particle P_1 again goes to its previous location. But it is NOT. Why? The reason has explained in the page 31-34 of this research book.

Every particle should move together with other particles due to their kinetic energies absorbed from F_1. We apply Kinetic energy to P_1 only. There should be a time spending to pass the kinetic energy to the faraway particle from P_1. Therefore, all the particles do not absorb energies at once. Therefore, due to the absorbed kinetic energy differences by each particle, the size/shape/length of the object are not definite.

2.3.Conclusion of Chapter 02

Therefore, according to the definition of the "Distance travelled by an object" we can conclude the following fact:

After T=0, there was no any external force acting on the whole object. But after T=0, the distances travelling by the object during unit time intervals are not uniform. i.e. after T=0, although there was no any external force acing on the object, the object's moving velocity changes.

But Newton's first law of motion says: If there is no any external force acting on the object, the object should move with a constant velocity or that should be at rest.

Therefore it is obvious that **I have obtained a contradiction to the Newton's first law of motion.**

But when one single building matter particle hasn't any external force acting on it, it must move with a constant velocity (Because, there is no any separate motions of the single fundamental matter particle).

i.e. Newton's first law of motion should be corrected.

Absolutely, a scientist can check my argument experimentally. He should build a vacuum with free space inside it. Here, "Free Space" is a space that has not been affected by any of the fundamental forces of nature. But, how he builds a space-time without any external forces???

Initially, he should consider the gravitational waves inside the experimental frame of box. In order to make that, he has to use exactly same amounts of matter and anti-matter along the two opposite sides. By using those exactly same amounts of matter and anti-matter, he will be able to create space-time without any net gravitational force acting on that space.

He must use the similar procedures to create the vacuum without any electromagnetic net force or net weak force. Strong force is limited to short range. Therefore he does not want to worry about the making procedure of the empty space inside the laboratory.

Then just keep a sensitive shining object inside the experimental vacuum box. At T=0, apply some non- zero pushing force. By using slowly moving instrument (Which is moving along the direction of applied force) it is capable to conclude whether the object is contracting and expanding time to time or not (By considering the photons those are coming towards the instrument from a particular point on the object, it is capable of measuring the locations of that particular point during each unit time intervals).

If the whole volume size of an atom cannot be changed in anyway, then P_1 absorbs E_1 energy and moves along the direction of motion without changing the electron orbital sizes. Also if we supply a sufficient energy to the P_1 atom, its electrons change the quantum state levels by exciting (Such as a pushing force with some heat). Also the nucleus of the atom P_1 absorbs some amount of energy from the applied external force. Then the quantum levels of the nucleons changes (The change of quantum levels will more obvious if we apply some pushing force by using thermal radiations and etc).

Under these excitations of electrons, protons and neutrons, the principle quantum level, orbital quantum level, orbital magnetic quantum level, spin quantum level, spin magnetic quantum level can be changed.

In the real situation,

$$E_1 = E_K + \sum_1^n E(i,p) + \sum_1^n E(i,l) + \sum_1^n E(i,ml) + \sum_1^n E(i,s) + \sum_1^n E(i,ms) + \sum_1^n E(i,os) + E_2 + E_3 \dots\dots \textbf{2.1}$$

Where, E_1 is the energy absorbed by P_1 from F_1 to move Δd distance away from the point A, along the direction of motion. E_K is the energy used to travel Δd distance by P_1. E_2 is the energy that passed to P_2.
E_3 is the energy P_1 has after it moved Δd distance.

$\sum_1^n E(i,p)$: Total Energy used to increase the principal quantum level of electrons and nucleons.

$\sum_1^n E(i,l)$: Total Energy used to increase the orbital quantum level of electrons and nucleons

$\sum_1^n E(i,ml)$: Total Energy used to increase the orbital magnetic quantum level of electrons and nucleons

$\sum_1^n E(i,s)$: Total Energy used to increase the spin quantum level of electrons and nucleons

$\sum_1^n E(i,ms)$: Total Energy used to increase the spin magnetic quantum level of electrons and nucleons

$\sum_1^n E(i,os)$: Total Energy spent due to the spin-orbit interaction of electrons and nucleons

Where n is the number of fermions in the atom P_1. Where "i" is the fermion order number in the atom P_1.

Due to the changes of the quantum states of fermions in the atom P_1, there is a sudden change of the forces among the fermions of P_1 and between P_1 and P_2. Let's find the new electromagnetic force among fermions of P_1: E(1,2).

$$E(1,2)=\Sigma Ei,j(e,e)+\Sigma Ei,j(e,p)+\Sigma Ei,j(e,n)+\Sigma Ei,j(p,p)+\Sigma Ei,j(p,n)+\Sigma E i,j (n,n)....2.2$$

Where $E_{i,j}$ (e,p) = q_i .Q_j / $4\pi\varepsilon r^2$ = the electromagnetic force between ith electron and jth proton of P_1 and similar symbolizations for other fermions.

Now let's find the **new strong nuclear force** inside the nucleus of P_1, E'(1,2).

$$E'(1,2)=\sum_{j=1}^k \{\sum_{i=1}^l E'(ni,pj)\}+\sum_{j=1}^k \sum_{i=1}^k E'(pi,pj)+ \sum_{j=1}^l \sum_{i=1}^l E'(ni,nj)...............2.3$$

Where E'(ni, pj) = The strong nuclear force between ith neutron and the jth proton

E'(pi, pj) = the strong nuclear force between ith proton and the jth proton

E'(ni, nj) = The strong nuclear force between ith neutron and the jth neutron

k is the # protons in the nucleus of P_1, l is the # neutrons inside the nucleus of P_1. Each term in (2.3) depend on the spin changes and orbital momentum changes of the nucleus of P_1.

Strong nuclear force is a short range nuclear force. Therefore, there is no an observable strong nuclear force acting between nucleus of P_1 and nucleus of P_2. But there is a change of electromagnetic force acing between P_1 and P_2 due to the changes of spin energy quantum levels and the orbital momentum energy quantum levels of P_1.

Let's denote new electromagnetic force acting among the fermions of P_1 and P_2 by E"(1,2).

E"(1,2)=ΣE" i,j (e,e)+ΣE" i,j (e,p)+ΣE" i,j (e,n)+ΣE" i,j (p,p)+ΣE" i,j (p,n)+

ΣE" i,j (n,n)....2.4

Where $E"_{i,j}$ (e,p) = $q_i \cdot Q_j / 4\pi\varepsilon(r")^2$ = The electromagnetic force between i th electron of P_1 and j th proton of P_2 and similar symbolizations for other fermions of P_1 and P_2.

Depending of the sign of E"(1,2) , E'(1,2) and E(1,2); the total distance traveled by P_1 until the situation 03 (in the page 26) should vary.

Because if the sign of E"(1,2) is an attractive force, then P_1 should further move along the direction of motion rather than Δd. But if E"(1,2) is a repulsive force, then P_1 can't come to the initial position that was at T=0. Why? Because, if P_1 comes to the initial position again due to the repulsive force of E"(1,2), then by (2.1) and (2.4); **$E_K = E''(1,2)*\Delta d$.**

Then **E_1 - $(\Sigma\ E)$ – E_2 –E_3 =** $\Sigma\ \Delta d*E''_{i,j}(e,e) + \Sigma\Delta d*E''_{i,j}(e,p) + \Sigma\Delta d*$ $E''_{i,j}(e,n) + \Sigma\ \Delta d*E''_{i,j}(p,p) +$ $\Sigma\ \Delta d*E''_{i,j}(p,n) + \Sigma\Delta d*\ E''_{i,j}(n,n)$**....2.5**

Where $E_3 = E_0(1,2)+E_0'(1,2) +E_X$; Where E_X is kinetic energy that the particle P_1 has traveled after travelling Δd.

Where $E_0(1,2)$ = new electromagnetic energy among the fermions of P_1, after changed its fermion's quantum levels

$E_0'(1,2)$= new strong nuclear energy among the fermions of P_1, after changed its fermion's quantum levels

But $\Sigma\Delta d*E''_{i,j}(e,e)$, $\Sigma\Delta d*E''_{i,j}(e,p)$, $\Sigma\Delta d*\ E''_{i,j}(e,n)$, $\Sigma\Delta d*E''_{i,j}(p,p)$,

$\Sigma\Delta d*E''_{i,j}(p,n)$ and $\Sigma\Delta d*\ E''_{i,j}(n,n)$ have magnitudes of 10^{-X}; where x is a

natural number.

Usually we apply F_1 = 1N or like that as F_1. It is not enough to apply 10^{-X} ordered pushing force to a considerably large object. But E_K is the kinetic energy used by P_1 to travel Δd distance. But after we apply F_1, the whole object should move within very small time gap. Therefore, the velocity of P_1 should be a high value. Although the mass of P_1 is a small value, due to high velocity of P_1, the energy E_K cannot be a small value as $\Sigma\Delta d*E''_{i,j}(e,e)$ and other terms in (2.5). Therefore the situation in (2.5) cannot happens.

Therefore, due to the repulsive force, P_1 can't come to the initial position that was at T=0 (As in situation number 03).

After that, energy amount of E_2 absorbs by P_2. Then we can write similar equations for P_2 by considering the previous procedures. We can write similar equations as (2.5) for all other particles.

Therefore, after some particle moved some distance due to the energy absorbed from F_1, they do not come to their initial positions due to the repulsive electromagnetic forces from other surrounding particles.

But in the atomic levels, the gravitational forces are negligible. Weak forces do not apply to this particular situation. Also strong nuclear force is limited to very small area (i.e. nucleus) and does not appear across two atoms.

Therefore, the only considerable force acting between two atoms is the electromagnetic force. As above details, after P_1 moved distance Δd, it does not come to its initial position. And the same incident for all other particles.

Here we can observe that although P_1 has moved Δd distance along the direction of motion, all other particles are at rest. In the 2nd step, we can observe that although P_1 and P_2 are moving along the direction of motion all other particles are at rest.

Therefore the situations described by situation # 1, situation # 2, situation # 3 and situation # 4 all are real.

If someone says: All the particles have bounded together, and all particles move together; **Then it is okay**. Why? Then, that bounded object becomes our fundamental particle inside the block of matter. But in order to disprove the Newton's first law of motion, I have used an example, i.e. an object that contains more than one fundamental particle. Our fundamental particle is: A particle with fixed dimensions and cannot change the size of the particle under any influences of an external force.

Conclusion

After T=0, there was no any external force acting on the whole object. But after T=0, the distances travelling by the object during unit time intervals are not uniform. That means, after T=0, although there was no any external force acing on the object, the object's moving velocity changes.

Therefore, I have ended up with the result:

The "NEWTON'S FIRST LAW OF MOTION IS NOT REAL"

Photo origin: science.pppst.com

Chapter 03

Discrete time flowing and a contradiction to Einstein's assumption

3.1. Introduction to the foundation for the innovative concept

Science and Technology developed through the mind-based concepts, new notions and the experimental procedures. But the current understanding of the natures of the physical quantities in the universe may or may not be the correct understanding due to the affect of extreme situation variations.

The real situation of the physical world may be different from our expecting world. In our usual physical world, we can observe so many incidents which are simultaneously happening. But we have to pay our attention to the real state of our basic concepts of some basic physical quantities. My attempt would use to realize a new idea regarding the "our understanding of the quantity time".

Due to the limited velocity of light, there is a time duration to propagate a light ray between any two distinct space-time points in the universe. Upon that argument, author's attempt is to obtain a very specific result that may useful for the Cosmology subject fields, string theory and the Astronomy subject fields. **The concept in this research article may be a challenge to almost all Physics subject areas.**

The final result implies that the time flowing depends on the space time location and time flowing is a relative fact in the universe. "Time flowing is a relative fact" does not mean the notion in the Special theory of Relativity regarding the relativity of the time.

3.2. Basic equations and concepts

We consider two persons call A and B with moving objects in their hands. My attempt is to consider A and B for 3 different situations called as Case 1, Case 2, Case 3. Consider each case separately as below.

Case A (person a and person b both are at rest relative to the frame)

Let assume, at point A and B there are two persons (those two persons consider as two points) carrying two simple clocks which are set into 00.00 am. And let assume, at time 't', both A and B can see himself **and** other both are moving their own pencils. (i.e. 'A' can see that him-self and 'B' both are moving their own pencils. And for 'B', 'B' can see himself and 'A' both are moving their own pencils at time 't')

That means A can see both two incidents (Him-self moves his pencil and he observes B also moves his pencil) simultaneously. Also the same capability for 'B'.

And $t=t_1$ is the time which is, according to the 'A''s clock whenever A is moving his pencil. And $t=t_2$ is the time which is, according to the 'B''s clock whenever 'B' is moving his own pencil.

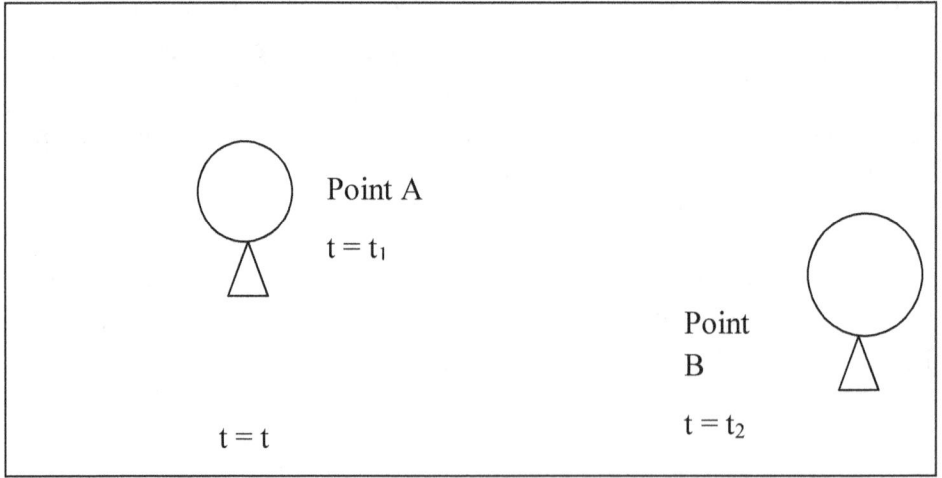

Figure 3.1 (within the same reference frame)

But at time t, 'A' can see himself and 'B' both are moving their pencils. Therefore $t > t_2$

$$t > t_2 \dots\dots\dots\dots\dots\dots\dots\dots\dots\dots\dots\dots\dots\dots\dots\dots\dots\dots(01)$$

(Because the light ray that is coming from B, spends some time on its path of travelling. But when time equals to t, A sees that B is moving his pencil. Then 'A' thinks 'B' had moved his pencil some bit before)

And also at time t, A knows that 'B' can see himself and 'A' both are moving their own pencils.

Therefore,

$$t > t_1 \dots\dots\dots\dots\dots\dots\dots\dots\dots\dots\dots\dots\dots\dots\dots\dots\dots\dots(02)$$

(Because the light ray coming from A spend some time on its path of travelling. But when time equals to t, A knows that B sees that 'A' is moving his pencil. Then 'B' thinks 'A' had moved his pencil some bit before)

By the equation (01) t is the time such that 'A' moves his own pencil.

Therefore by the equation (01), $\quad t_1 > t_2 \dots\dots\dots\dots\dots\dots\dots\dots\dots\dots\dots(03)$

By the equation (02) t is the time such that B moves his own pencil.

Therefore, by the equation (02), $t_2 > t_1 \dots\dots\dots\dots\dots\dots\dots\dots\dots\dots(04)$

Therefore by the equations (03) and (04) (which are A's conclusions), we get a paradox. Therefore we can conclude that our previous assumption is false.

Therefore, **NO ONE** can see two incidents (Two incidents: one is done by himself and other done by another person) such that those are **simultaneously happenings.**

Case B (Both A and B are in the same inertial frame, but A is moving with Velocity V ms-1 Relative to the person B)

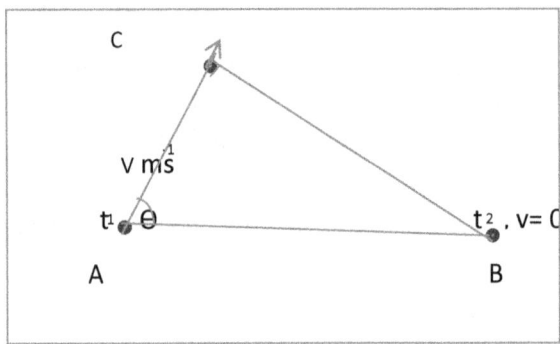

Figure 3.2 (within the same inertial frame)

Consider the case which is: The person at the position A, moves with constant velocity V ms^{-1} with Θ angle ($0 \leq \Theta < 2\pi$) with respect to the line AB. And B is at rest within the frame. While moving with velocity V, A sets his clock to 00.00am. At the same moment B also sets his clock to 00.00am. Then start to consider next incidents. **i.e. the below t_1 and t_2 indicate the time (t_1 – 00.00) and time (t_2 – 00.00). For case B, (t_1 -0.00) = the time duration taken by A, to move his own pencil (According to A's own clock). (t_2 – 0.00) = time taken by B to move his own pencil (According to B's own clock)**

Where $t = t_2$ is the time according to B's clock (which is at rest within the frame) whenever B moves his pencil. And $t = t_1$ is the time according to the A's clock (which is moving with velocity V ms^{-1} with respect to the person B) whenever A moves his own pencil. **Let assume at time ' t ' , B can see himself and other both are moving their own pencils(i.e. at time ' t ' according to the B's clock). And also assume B knows that at time 't' (According to B's clock) ; A can see himself and B both are moving their pencils.**

Let, time $= t_m$ is the time in B's clock when A moves his pencil. Therefore, $t_1 / \sqrt{1- (V/C)^2} = t_m$ (Because t_1 is the improper time for B.

Then B says that $t > t_m$ (because light ray coming from A to B spends some time on its path of travelling and at time t in B's clock, B sees that A is moving his pencil)

$t > t_1 / \sqrt{(1- (V\backslash C)^2)}$……………………. 05'(because t_1 is the improper time for person B. In relativity, improper time is time measured by a single clock between events that occur at the different place as the clock)

But at time 't', B knows that himself is also moving his pencil. Therefore, B says that $t = t_2$

Therefore by (05'), $t_2 > t_1 / \sqrt{(1- (V\backslash C)^2)}$…………………………..(05)

Now let's consider the light ray propagating from B to A. But **B knows** that at time t, **A receives the light ray coming from B** (Because at time t according to B's clock, B knows that both A and B can see himself and other both are moving their own pencils. i.e. at time ' t ' according to the B's clock). Then B sees the time of B's clock when A is receiving the light ray (came from B) as $t = t_1 / \sqrt{1- (V/C)^2}$. Because the time t_1 according to A's clock is the time that A moves his pencil. (Because at time t_1 according to A's clock, A receives the light ray coming from B and 'A' moves the pencil both happens). Therefore,

$t_1 / \sqrt{1- (V/C)^2} = t = t_2$ …………………………………………………..(06)

By the equations 05 and 06 we get a contradiction. Therefore our assumption is false. Therefore such a time ' t ' does not exist. ………………(Result N)

In the above case I didn't assume that A and B moves their pencils at the same time. Now assume 'A' and 'B' moves their pencils at the same time.

Let's consider, $t_2 / \sqrt{(1 - (V\backslash C)^2)} = t_1$ (i.e. A knows whenever he moves his pencil B is also moving his pencil). Because $t_2 / \sqrt{(1 - (V\backslash C)^2)}$ is the improper time for A and that is same as t_1.

(Because in relativity, improper time is time measured by a single clock between events that occur at the different place as the clock).

And for person at B, we get the similar statement $t_1 / \sqrt{(1 - (V\backslash C)^2)} = t_2$. (i.e. B knows whenever he moves his pencil A is also moving his pencil). That means both A and B moves their own pencils at the same time.

Then by considering person at B, there is a time t =

$t_1 / \sqrt{(1 - (V\backslash C)^2)}$ ....(K)

And by the statement $t_1 / \sqrt{(1 - (V\backslash C)^2)} = t_2; t = t_2$; i.e. t is the proper time for 'B'. But B is moving his pencil at time t_2 according to his own clock. Then the time $t = t_2$, as A is seeing according to A's clock = $t_2 / \sqrt{(1 - (V/C)^2)} = t_1 / (1 - (V\backslash C)^2)$ (By using (K) and since t_2 is an improper time for A).

Then the time spending **(T'')** to travel the light ray from A to B as A is seeing = $(t - t_1) = [t_1 / (1 - (V/C)^2)] - t_1] > 0$. Author can write the time that the light ray from A to B starts to propagate from A as t_1. Then the time taken to propagate the light ray from A to B as A is detecting as same as the time taken to propagate from A to B as B is detecting. Because the time taken by the light ray to propagate a constant distance is independent from the observer.**(But in order to receive the light ray by B, B should be [T''.C] distance away from A. C is the speed of light)** Then there is a time 't' such that person 'B' can conclude that B is moving his own pencil and B can see the movement at A at the same time t.

Because there is a definite chance to receive the light ray by B coming from A, if A and B are   [T''.C] distance away from each other.

Then by considering person at A, there is a time

$t = t_2 / \sqrt{(1 - (V/C)^2)}$ ....(K')

And $t = t_2 / \sqrt{(1 - (V/C)^2)} = t_1$ is the time in A's clock when A moves his pencil. Then the time $t_1 (= t_2 / \sqrt{(1 - (V\backslash C)^2)}$, as B is seeing according to B's clock = $t_2 / (1 - (V/C)^2)$

(By using (K') and since t_1 is an improper time for B).

Then the time taken (T''') to travel the light ray from B to A, as B is seeing (Then the time taken to propagate the light ray from A to B as A is detecting as same as the time taken to propagate from A to B as B is detecting.) = $[t_2 / (1 - (V/C)^2) - t_2] > 0$. Because the time takes by the light ray to propagate a constant distance is independent from the observer. **(But in order to receive the light ray by A, A should be [T'''.C] distance away from B. C is the speed of light)**

Then there is a time '$t = t_2 / \sqrt{(1 - (V/C)^2)} = t_1$' in A's clock such that person 'A' can conclude that A is moving his own pencil and A can see the movement at B at the same time $t = t_2 / \sqrt{(1 - (V/C)^2)} = t_1$ ....(**)

But **B can conclude** that at time $t_1 / \sqrt{(1 - (V/C)^2)}$, (according to B's clock) **A knows that** A is moving his pencil and A can see movement at B at the same time(Because t_1 is an improper time for B). And also, at time '$t = t_1 / \sqrt{(1 - (V/C)^2)} = t_2$ in B's clock, person 'B' can conclude that B is moving his own pencil and B can see the movement at A at the same time $t = t_1 / \sqrt{(1 - (V/C)^2)}$ ....(Paragraph M)

But in order to conclude that A and B are should be T_k distance away from each other. Where $T_k = \min \{T'', T'''\}$

But we know that as paragraph M, such a time 't' does not exist. (According to the result N). Then we get a contradiction.

Therefore our assumption is false. Therefore, we can conclude that 'A' and 'B' can't move their pencils at the same time. Therefore, in the case B we can conclude that 'A' and 'B' can't move their pencils at the same time if they are in the distance [T_k .C] away from each other.

Where C is the speed of light and $T_k = \min \{T'', T'''\}$

Case C: consider the case person at A, moves with acceleration (this acceleration may or may not be a constant) with respect to the person at B.

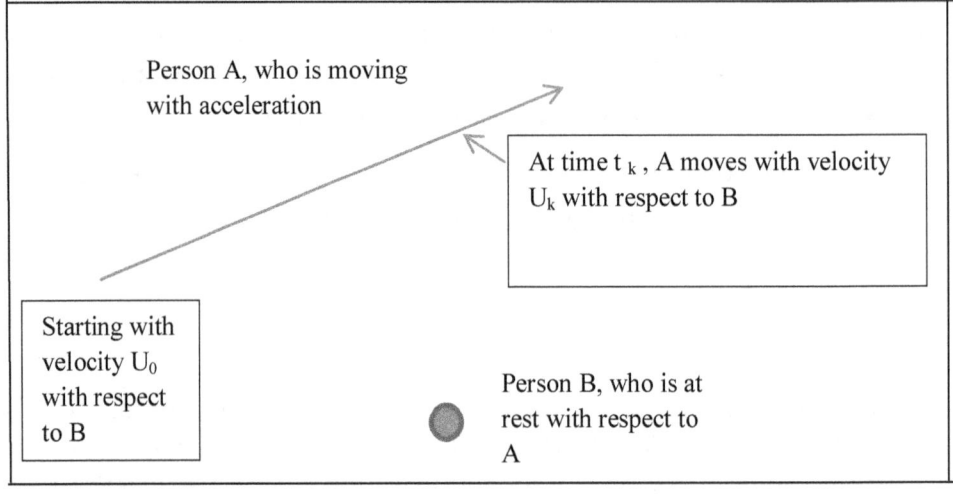

Person A, who is moving with acceleration

At time t_k , A moves with velocity U_k with respect to B

Starting with velocity U_0 with respect to B

Person B, who is at rest with respect to A

Figure 3.3: person at A, moves with acceleration (this acceleration may or may not be a constant) with respect to the person at B.

At point A, the velocity of that person is U_0 ms^{-1} with respect to the person at B. And at $t = t_k$, the velocity of the person (person: who was at A) is U_K ms^{-1}. While moving with acceleration, A sets his clock to 00.00am. At the same moment B also sets his clock to 00.00am. Then start to consider next incidents. **i.e. the below t_0 and t' indicate the time (t_0 – 00.00) and time (t' – 00.00). For case C, (t_0 -0.00) = the time duration taken by A, to move his own pencil (According to A's own clock). (t' – 0.00) = time taken by B to move his own pencil (According to B's own clock)**

And $t = t_0$ is the time according to the clock of the person who was at point A when he moves his pencil ($t = t_0$ is the time according to the (moving)A's clock). And $t = t'$ is the time according to B's clock at position B, when B moves his pencil. Let assume at time ' t ' , B can see himself and other both are moving their own pencils(i.e. at time ' t ' according to the B's clock). And B knows that at time 't'(according to B's clock) ; A can see himself and B both are moving their pencils.

By considering the light ray coming from A to B, 'B' thinks that,

$$t' > t_0 / (1 - (U_0 \backslash C)^2) \dots\dots\dots\dots\dots\dots\dots\dots\dots\dots\dots 07$$

Same procedure as case B

By considering the light ray coming from B to A , B thinks that ,

$$t_0 / \sqrt{(1 - (U_0 \backslash C)^2)} = t' \dots\dots\dots\dots\dots\dots\dots\dots\dots\dots 08$$

By equation (07) and (08) we get a contradiction. Therefore, there does not exist such a time 't'.

Let's consider, $t' / \sqrt{(1 - (V \backslash C)^2)} = t_0$ (i.e. A knows whenever he moves his pencil B is also moving his pencil). Because $t'/ \sqrt{(1 - (V \backslash C)^2)}$ is the improper time for A and that is same as t_0. (Because in relativity, improper time is time measured by a single clock between events that occur at the different place as the clock)

And for person at B, we get the similar statement $t_0 / \sqrt{(1 - (V \backslash C)^2)} = t'$. (i.e. B knows whenever he moves his pencil A is also moving his pencil). That means both A and B moves their own pencils at the same time.

Then by considering person at B, there is a time $t = t_0 / \sqrt{(1 - (V/C)^2}$

Then, $t = t'$. (by the statement $t_0 / \sqrt{(1 - (V \backslash C)^2)} = t'$).

Then the time spending (T_a) the light ray on the path of travelling from A to B as A is seeing = $(t - t_0) = [t_0 / (1 - (V/C)^2)] - t_0] > 0$. Author can write the time that the light ray from A to B starts to propagate from A as t_0. Then the time taken by the light ray to propagate from A to B as A is detecting as same as the time taken by the light ray to propagate from A to B as B is detecting. Because the time taken by the light ray to propagate a constant distance is independent from the observer **(But in order to receive the light ray by B, B should be [T_a .C] distance away from A. C is the speed of light).** Then there is a time $t = t_0 / \sqrt{(1 - (V/C)^2}$ according to B's clock such that person 'B' can conclude that B is moving his own pencil and B can see the movement at A at the same time t. ....(P)

Then by considering person at A, there is a time $t = t' / \sqrt{(1 - (V/C)^2)}$. And $t = t' / \sqrt{(1 - (V/C)^2)} = t_0$ is the time in A's clock when A moves his pencil.

Then the time $t_0 (= t' / \sqrt{(1 - (V\backslash C)^2)}$, as B is seeing according to B's clock $= t'/ (1 - (V/C)^2)$ (Similar to the procedure in case B)

Then the time taken (T_b) to travel the light ray from B to A, as B is seeing (Then the time taken to propagate the light ray from A to B as A is detecting as same as the time taken to propagate from A to B as B is detecting.) $= [t'/ (1 - (V/C)^2) - t'] > 0$. Because the time takes by the light ray to propagate a constant distance is independent from the observer. **(But in order to receive the light ray by A, A should be [T_b.C] distance away from B. C is the speed of light)**

Then there is a time $t = t' / \sqrt{(1 - (V/C)^2)} = t_0$ in A's clock such that person 'A' can conclude that A is moving his own pencil and A can see the movement at B at the same time $t = t' / \sqrt{(1 - (V/C)^2)} = t_0$ ....(**)

But **B can conclude** that at time $t_0/ \sqrt{(1 - (V/C)^2)}$, (according to B's clock) **A knows that** A is moving his pencil and A can see movement at B at the same time (Because t_0 is an improper time for B). And also, at time '$t = t_0 / \sqrt{(1- (V/C)^2)} = t$' in B's clock, person 'B' can conclude that B is moving his own pencil and B can see the movement at A at the same time $t = t_0 /\sqrt{(1- (V/C)^2)}$ ....(P)

Then by (P) and (**); we can conclude that there is a time $t_0 / \sqrt{(1- (V/C)^2)}$ in B's clock such that 'B' can conclude that at time $t_0 / \sqrt{(1- (V/C)^2)}$ (according to B's clock),**A knows that** A is moving his pencil and A can see movement at B at the same time. And also, at time $t = t_0 / \sqrt{(1- (V/C)^2)}$ in B's clock, person 'B' can conclude that A is moving his own pencil and B can see the movement at A at the same time $t = t_0 / \sqrt{(1- (V/C)^2)}$ ....(Paragraph Q)

But according to the Case C initial contents, author knows such a time t does not exist (But A and B should be [T_c .C] distance away from each other. Where $T_c = \min \{ T_a, T_b \}$.)

Then we get a contradiction. Therefore 'A' and 'B' can't move their pencils at the same time if they are in [T_c .C] distance away from each other. Therefore, in the case C, we can conclude that 'A' and 'B' can't move their pencils at the same time if they are in [T_c .C] distance away from each other.

A and B can't move their own pencils at the same time if A and B in the distance [T_k.C] and [T_c .C] away from each other respectively for case 2 and case 3. THEREFORE IN 'CASE B' AND 'CASE C' NO TWO PERSONS CAN DO TWO WORKS SIMULTANEOUSLY.

Then the possible implications are,

(For two persons such that they are in relative motion and T_k ,T_c distance away from each other for the case 2 and case 3 respectively)

1. The happening time for an incident creates with that incident and dies with that incident(that means time births with that incident and dies with that incident: time is just like a shadow of a person)
2. Time is flowing in a discrete manner(that means time is not flowing without stopping)

But if (2) is true, then there can be two incidents those are happening at the same time. But , it cannot happens. Therefore 1st implication is true.

That means time is related with that incident if they are in [T_k .C] distance OR [T_c .C] distance away from each other respectively.

We know that universe filled with matters and energy. Let us consider the physical term **" velocity "** . First consider the velocity of an Energy wave (that means a movement without any matters).

Case 1

Consider a movement of a wave(without any matter where , that wave consider with respect to a matter's motion OR with respect to another wave's motion. i.e. there should be a relative motion with respect to the first wave. Let's consider an electromagnetic wave. **(The velocity of that electromagnetic wave should be smaller than 'C' during the consideration period: due to some reasons)**Then the velocity of that EM wave defines as,

Velocity =   distance travelled that EM wave within a unit time

Velocity = $\Delta d / \Delta t$

But now we know that a time births and vanishes with the incident. **(Time births with the incident and has been died by the influence of the incident- then only we can say two incidents haven't occurred at the same time).** With this new idea of time we can write the quantity Δt as,

$\sum_1^\alpha \Delta t_i = \Delta t;$ α denotes infinity.

Here, ' i ' is the incident number.

Δt_i is the time that births with the incident 'i' . And

Δt_i dies due to the influence of incident 'i'.

Here **'incident' is the smallest possible distance (that can travel) travelling without giving birth to another time** Δt_{i+1}**. Moreover,** Δt_i is the time interval which does not allow to happen 2 incidents simultaneously (that does not allow to happen at least two incidents)

And Δt_i = time taken by the wave in order to travel Δd_i ($\Delta d_i \rightarrow 0$) distance. Here, Δt_i should be the time interval that does allow to happen only one incident. Because with a new incident, another time births. Δd_i **is the smallest possible distance that EM wave can travel within the time interval** Δt_i .

Explanation of the existence of smallest possible distance Δd_i **for the time interval** Δt_i **:**

Consider P and Q people. P is at the center O. Q is moving with the radius $R = T_k.C$ around P. If there is no smallest possible distance Δd_i that can travel within Δt_i ; we can say $\Delta d_i = \pi. R$

Then while Q is moving $\pi.R$ distance around P, the time duration that is flowing is Δt_i . Then we **cannot** say $\Delta t_i \rightarrow 0$. Then there may be an another incident that is happening at P. Because for large Δd_i , there is a large Δt_i (Comparably). Then within the same time duration Δt_i, there may be two incidents those are happening.

But we know that cannot happened. Therefore there is a smallest possible distance ($\Delta d_i \rightarrow 0$) that Q can travel within Δt_i.

And for all other situations and arrangements of two random people P and Q; we can conclude the same arguments.

These time birthing and vanishing procedure only for an incident that is $T_k.C$ distance or $T_c.C$ distance away from the other incident as previously mentioned) Where i ϵ N.

$\sum_1^\alpha \Delta ti = \Delta t$ should be a finite value. Therefore, we know $\sum_1^\alpha \Delta ti = \Delta t$ should converge to some value $k \epsilon R^+$.

Therefore with the help of mathematics we obtain:

$$\Delta t_{i+1} / \Delta t_i < 1; \qquad \Delta t_{i+1} < \Delta t_i \dots\dots\dots\dots\dots (09)$$

(09) implies that EM wave takes small time rather than it took before, in order to travel the distance $\Delta d_{i+1}(\Delta d_{i+1} \to 0)$.

Where,

Δd_i is the smallest possible distance that EM wave can travel within the time interval Δt_i .

But the velocity (V) of that EM wave is defined as , $V_i = \Delta d_i / \Delta t_i$ Where i ϵ N .

1.   If V_i is constant , then the value $\Delta d_{i+1} < \Delta d_i$ for i ϵ N . That means the smallest possible distance, that EM wave can travel within its own time Δt_i (because now we know time is just like shadow of a person) is smaller than it is before.

2.   **OR   If the smallest possible distance that EM wave can travel within the time interval Δt_i is a constant; then the velocity of EM wave should larger than it is before.**

$$V_i . \Delta t_i = \Delta d_i$$

$\sum Vi \cdot \Delta ti = \sum \Delta di$ $i \epsilon$ N .

But, $\sum \Delta di = d =$ distance travelled the EM wave within the whole considering period.

$\sum_1^\alpha Vi. \Delta ti = \Delta d$...................(10)   Where i ϵ N .

Case 2

Consider a movement of a matter. Then the velocity (U) of that matter is defined as:

$U_i = \Delta d_i / \Delta t_i$; i ϵ { 1, 2, 3,.................,n}

$d_i =$ smallest possible distance that the object can travel within the time interval Δt_i. Because, now we know that time is just like a shadow of a person and there is a smallest possible distance that can travel by an object during the time interval Δt_i. This Δt_i should birth whenever object starts to move d_i distance and should die after the object moved d_i distance.

Where n ϵ N . And n is finite. This n depends on the length of the line that particle travelled and the smallest possible distance that the particle can travel within our new time Δt_i ; i ϵ {1,2,3,.........,n}

Put $\Delta d_i = w_i$.

Then $U_i = w_i / \Delta t_i$

$w_i = U_i. \Delta t_i$

$\sum_1^n Ui. \Delta ti = \sum_1^n Wi$

$\sum_1^n Wi = L = \sum_1^n Ui. \Delta ti =$ distance travelled by the object within the considering period

$$L = \sum_1^n Ui. \, \Delta ti \qquad \dots\dots\dots\dots\dots\dots\dots\dots\dots\dots\dots(11)$$

* Specially, it should be noted that this definition valid only for movements such that there is a relative motion of the object/wave with respect to the observer.

Photo origin: www.youtube.com

3.3. Conclusions

Conclusion 1

But this discrete time flowing definition is valid for the objects who are in a specific distance away from each other. That means the **'TIME FLOWING DEPENDS ON THE SPACE'.**

Let's consider two persons P and Q. P is at rest with respect to the rest space time. And Q is moving with velocity 'V' in a circular path around P. Then for some particular radius R', around the person P, the relative time flowing of Q is discrete with respect to P, according to above all arguments.

But P and Q are in different inertial frames if Q moves around P.

THEREFORE EINSTEIN'S STATEMENT: "ALL PHYSICAL LAWS ARE SAME FOR ALL INERTIAL REFERENCE FRAMES" MAY HAS A PARADOX, ACCORDING TO ABOVE ARGUMENT.

And I hope to develop the main idea in this research article further, and it will be able to explain some unexplained Astrophysics, Physics and Cosmology problems in future.

Conclusion 2

Let's consider two persons X and Y inside a same inertial frame S. X is at rest relative to the frame S. And Y is moving with velocity V' with respect to X within the inertial frame. Y is moving along a straight line starting from minus infinity (Comparably) to plus infinity (Comparably).

And also, the shortest distance (D) to the person X from the straight line should change time to time.

Then my previous works imply that there should be at least one situation (with distance D = D') of the straight line such that the time does not flow continuously with Y, with respect to X (On at least one spot along the straight line – straight line which is distance D' away from the person X).

But there may be more than above particular situation (Particular situation: With a straight line D' distance away from X and at the particular point on that straight line that previously considered

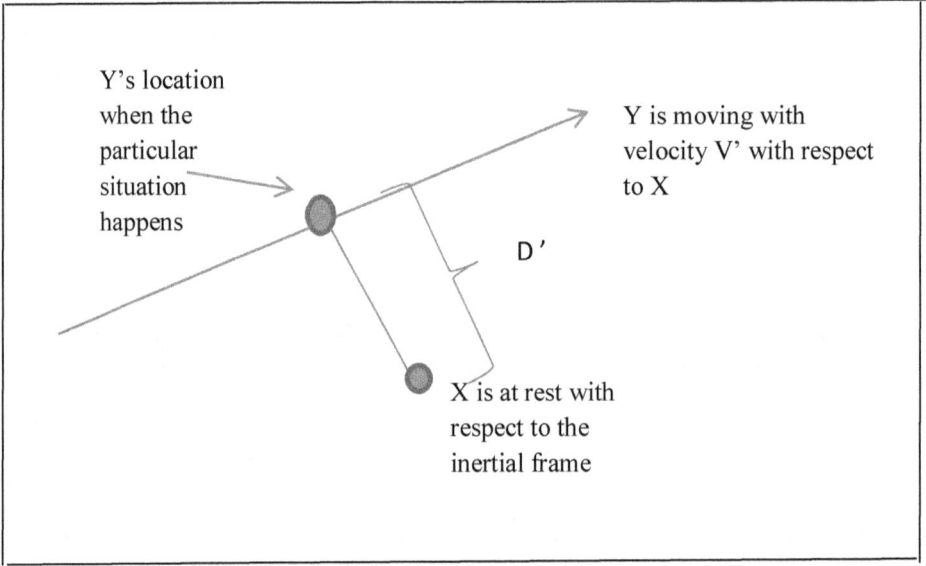

Y's location when the particular situation happens

Y is moving with velocity V' with respect to X

D'

X is at rest with respect to the inertial frame

Figure 3.4: Within the same inertial frame (When the time flowing of Y becomes discrete)

Therefore although the time of X flows continuously, the time flowing with Y is not continuous (For the above particular situation) within the same inertial frame. Therefore, although the two persons X and Y are in the same inertial frame, the laws of physical quantity 'time' are different from each other (Because for X, time flows continuously, but not for Y).

But Einstein has considered that the laws of physics are same for all inertial frames.

But according to my above proofs, the laws of the physical quantity 'time' are not same within the same inertial frame either.

before   after

Photo origin: http://motls.blogspot.com/2007/02/resolving-einsteins-twin-paradox.html

Chapter 04

Detecting an asteroid

There were lots of collisions happened due to the asteroids coming from the outer space. Some incidents are very harmful for life on Earth. Now I try to investigate a new method to detect such asteroids coming from outer space and measure the velocity it moves. We know there is relative motion of Earth with respect to the asteroid; due to Earth's spinning and due to the revolution of Earth around sun. We know when an asteroid is coming towards the Earth, some energized electromagnetic waves should come towards the Earth. And we detect and measure those electromagnetic waves at two different Earth times. Then within that two different Earth times, Earth has gone some length along the orbit of the Earth as well as it has rotated some angle around the rotating axis of the Earth. But basically the asteroid moves in a straight path with respect to other stellar objects.

i.e. relative to sun. The situation is as below.

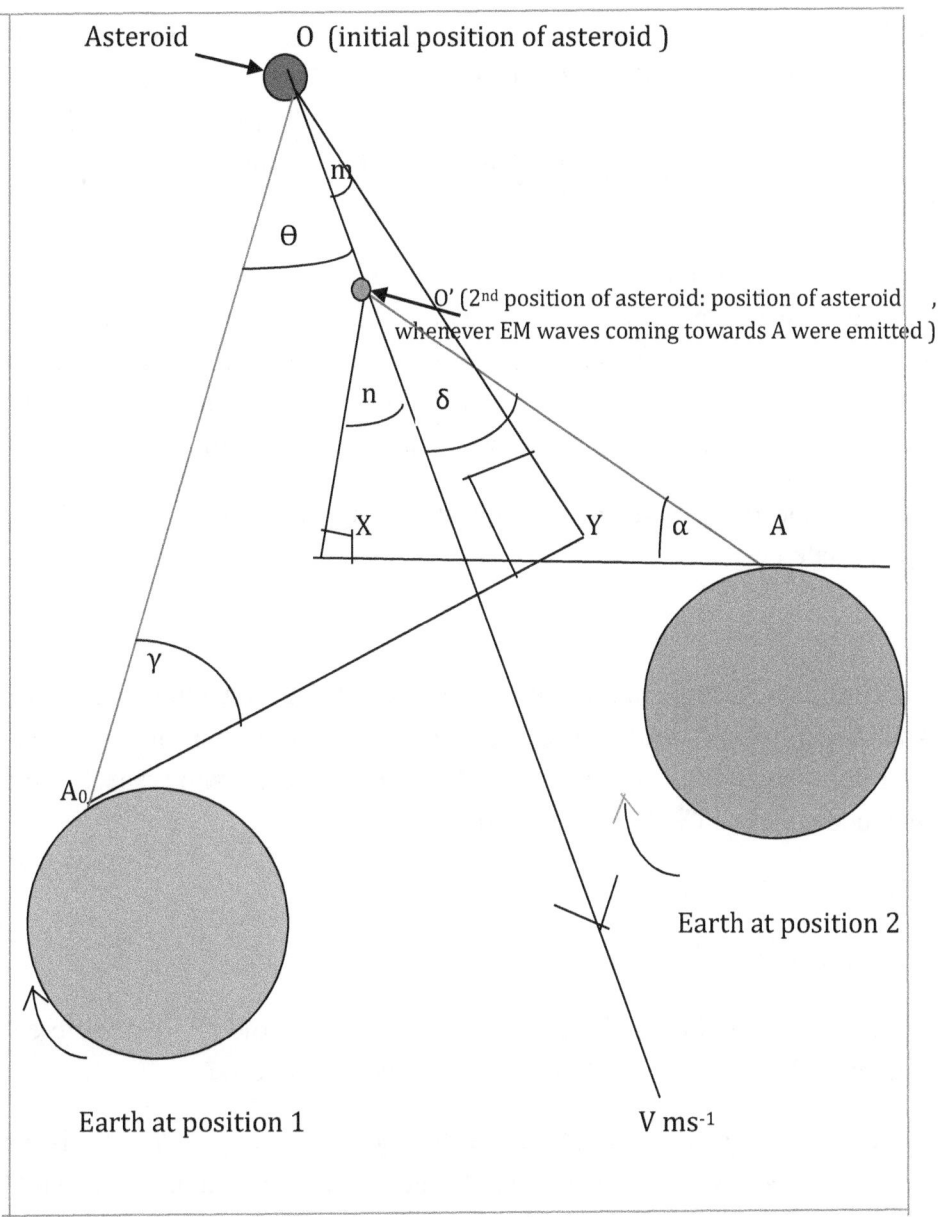

Figure 4.1

Where A_0 is the position that initial observation of the asteroid has done and A is the 2nd moment of asteroid detection and A and A_0 same position on Earth but at two different Earth times. And within that **time interval (t)** Earth has rotated β angle along its orbiting path around the Sun.

And θ is the angle between A_0O and asteroid coming direction. And γ is the angle between A_0O and tangent line to Earth at A_0 (A_0Y is the tangential line at point A_0 to the Earth and OY is the perpendicular line to A_0Y). And AX is the tangential line to Earth at point A and $O'X$ is the perpendicular line for line AX.

There are two motions of the Earth. Spinning and Revolving around the Sun. But if we keep the asteroid detector at the North Pole, then there is no any spinning angle of the asteroid detector, relative to an External observer. Here external observer is an observer who is in outside of the Earth's surface. He is in the deep space without any motion.

<u>i.e. If we locate the asteroid detector at North Pole of the Earth, then the angle rotated due to the Earth's spinning is zero.</u>

Let's neglect Earth's spinning angle for the asteroid detecting procedure. And Earth rotates 360 degrees around the sun within 365.25 days. Therefore the total angular velocity ($2\pi/T$) of the Earth with respect to the surrounding space (With respect to the person X)

$= 2\pi/(365.25*24)$

Therefore, $\beta = [(2\pi/(365.25*24))] * t$ ....(*)

; t is the Earth time difference of detecting asteroid at two different moments. β is the angle between the tangents at A_0 and A.

Let W is the tangential velocity of Earth at point A_0 along A_0Y, W' is the tangential velocity of Earth at point 'A' along the direction of A_0Y. And U is the velocity of the asteroid with respect to 'Earth detection' at the first moment of detection at point A_0. i.e. **U was the <u>previous velocity</u> of the asteroid along the direction A_0Y, <u>whenever now coming EM waves toward point A_0 were emitted.</u>** And U' is similar to the 2nd detection of EM waves at point A. And important: U is the velocity component of V at A_0 in the direction A_0Y and U' is also the velocity component of V, at position A, in the direction A_0Y.

Then $U = V \cos(\theta).\cos(\gamma)$............................(1)

<u>V is the velocity of the asteroid</u> with respect to the Earth, <u>whenever now coming</u> (now coming means: whenever the 1st detection EM waves coming from the asteroid reach point A_0) EM <u>waves were emitted before.</u>

$U' = - V'.\cos\delta.\cos \alpha.\cos \beta$. (minus include : because asteroid velocity component at the 2nd detection is in the opposite direction of A_0Y)

β is the angle between the tangents at A_0 and A.

V' is the velocity of the asteroid with respect to the Earth, whenever now coming (now coming means: whenever the 2nd detection EM waves coming from the asteroid reach point A) EM waves were emitted. i.e. V' is the past velocity of the asteroid.

We can consider

$V' = k.V$ k is a complex constant or function of V. Therefore

$U' = -kV.\cos\delta.\cos \alpha.\cos \beta$........................(2)

β explanation :

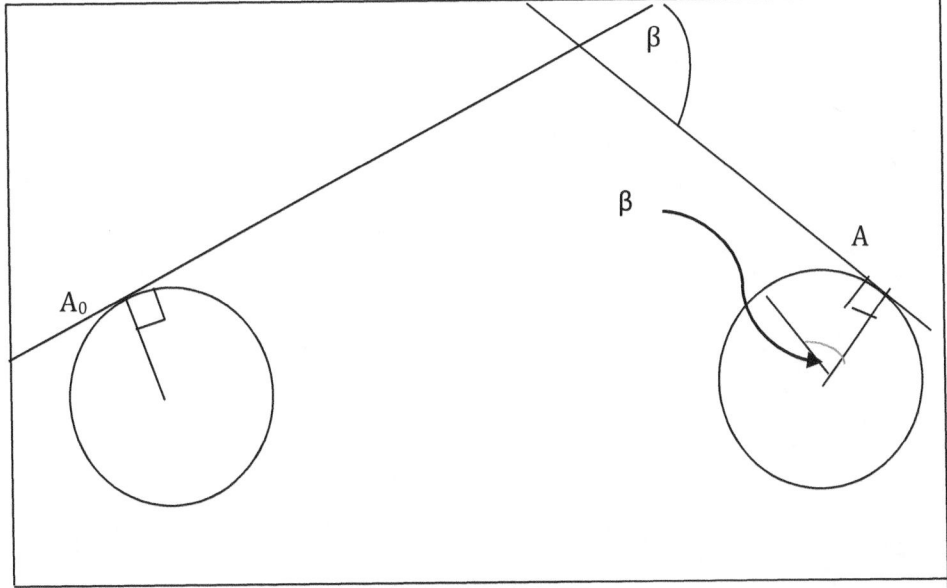

Figure 4.2

When Earth revolved β angle around its orbit of motion, the tangential line at point A_0 should also rotate an angle β. Therefore angle between A_0Y and AX is also β. Then the velocity component in the direction AX is equivalent to cosβ times that velocity component and that is in the direction A_0Y.

We know Earth has only one velocity component (If we keep detector at the North Pole). That is due to the Earth's revolution around the Sun.

Due to the revolution around the sun, there is a velocity component.

Revolution component of the Earth (along the direction A_0Y) = $r_0*(2\pi/365.25*24)*\cos\phi$ As shown in figure 4.3

Where $\phi = (2\pi*t_1)/24$

t_1 = time difference between the moment of asteroid's 1st detection occurs and the time of the place such that the sun is overhead (whenever 1st detection is occurring).

(Sun overhead position means: The spot on Earth where the detector and the Sun are in closest distance)

Explanation :

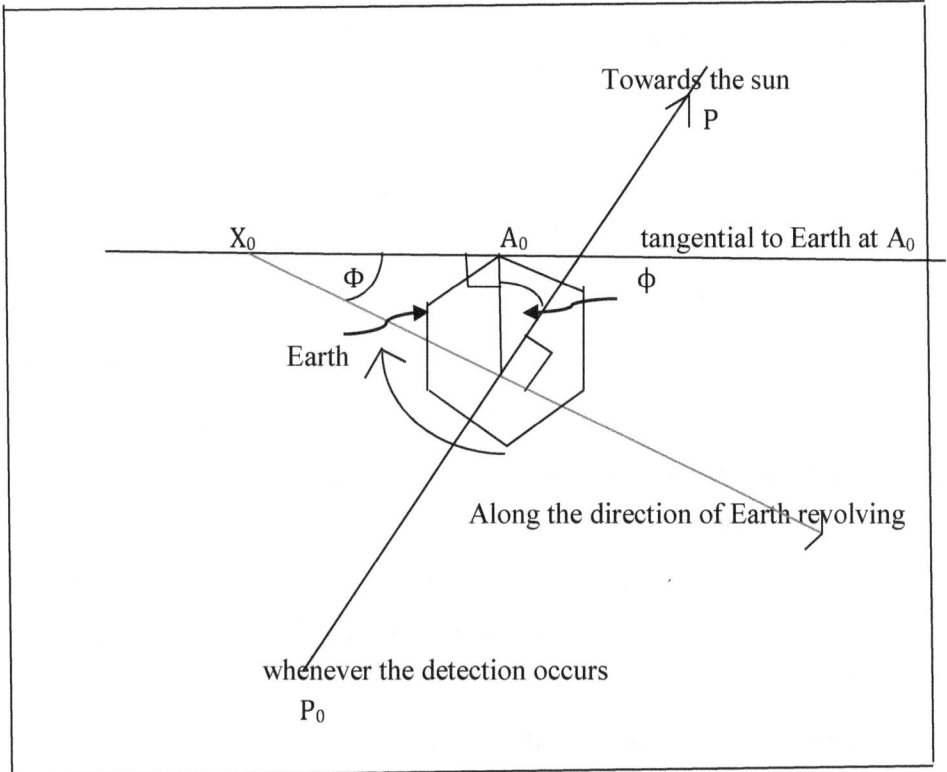

Figure 4.3

We know definitely PP_0 and XX_0 can be consider as perpendicular lines. Then we know that angle ϕ = (angular velocity * time taken to rotate the Earth about an angle ϕ). r_0 is the distance between Sun and Earth whenever the 1st detection occurs.

(Angular velocity of Earth to spin around its axis = $2\pi/T$)

time taken to rotate the Earth an angle ϕ = t_1 = time difference between the moment of asteroid 1st detection occurs and the time of the place such that the sun is overhead (whenever 1st detection is occurring) That (t_1) can be easily measure: just the time difference between two clocks placed at A_0 and sun overhead position (As in figure 4.3). (Sun overhead position means: The spot on Earth where the detector placed and the Sun are in closest distance)

Similarly, revolution component of velocity W' along the direction A_0Y =

r_0' * $(2\pi/365.25*24)*\cos \phi'$ *$\cos \beta$

Where r_0' is the distance between Sun and Earth whenever the 2nd detection occurs.

And $\phi' = (2\pi*t_2/24)$

Where t_2 is the time difference between the position of 2nd detection occurs and similar sun overhead position.

(Sun overhead position means: The spot on Earth where the detector located and the Sun are in closest distance)

Tangential velocity component of W due to the Earth's spinning = 0 (Because the detector has placed at the North Pole; where the spinning velocity is zero)

Tangential velocity component of W' along A_0Y due to the Earth's spinning = 0

Then according to the relationship ($V = r.\omega$)

W = [$r_0*(2\pi/365.25*24)*\cos \phi$]....................(3)

W'=[$r_0'*(2\pi/365.25*24)*\cos\phi'*\cos\beta$]....................(4)

We can measure the values of γ and α: By measuring the angle between the flat Earth surface at point A_0 and the direction of asteroid is observable; we can measure the angle γ. And by measuring the angle between the flat Earth surface at point A and the direction of asteroid is observable; we can directly measure the angle α.

We can measure the intensity of coming EM waves towards the point $A_0(I_1)$ by using suitable instrument as well as intensities of EM waves coming towards point A(I_2) We have,

When $\Theta \rightarrow 0$, $I_1 \rightarrow I_{1,max}$ and when $\delta \rightarrow 0$, $I_2 \rightarrow I_{2,max}$. Therefore using the equation of maximum intensity and the intensity equations for point A_0 and A, we can calculate the values of Θ and δ. Then we can get expressions for U and U' in terms of V. We can directly get the values of W and W' as constant values. Let $W = k_1$ and $W' = k_2$. Then the relative velocity of Earth with respect to the asteroid at position A_0 along the direction $A_0 Y =$

$$V_1 = k_1 - g_1(V)\dots\dots\dots\dots\dots\dots\dots\dots(5)$$

Then the relative velocity of Earth with respect to the asteroid, at position A along the direction $A_0 Y = V_2 = k_2 - g_2(V)\dots\dots\dots\dots\dots\dots\dots\dots(6)$

Where $g_1(V)$ is the expression get from equation (1) and $g_2(V)$ is the expression get from the equation (2). Then by relativistic Doppler formula we get

$$f_1 = [\ 1/\ \sqrt{(1-(V_1/C)\ )}\ ].\ \{\ 1-(W/U)\ \}.\ f_0$$

$$f_2 = [\ 1/\ \sqrt{(1-(V_2/C)\ )}\ ].\ \{\ 1-(W'/U')\ \}.\ f_0'$$

Where f_1 is the frequency of coming EM waves toward point A_0 those detected at point A_0. And f_2 is similar frequency at point A. And f_0 is the real frequency of EM waves coming towards A_0, when they were emitting. And f_0' is the real frequency of the EM waves coming towards A, when they were emitting by the asteroid. Since time difference between two detection of asteroid is very small, we can consider $f_0 = f_0'$. Therefore by the above equations,

$$f_1/f_2 = [1/\ \sqrt{(1-(V_1/C))}].\ \{\ 1-(W/U)\}/\ [\ 1/\sqrt{(1-(V_2/C))}\ ].\ \{\ 1-(W'/U')\ \}$$

Therefore we get the relation that contains V and k(V) terms. Therefore if we assume k(V) =1(that means velocity of the asteroid does not change within that time interval of two detections occurs), then we can estimate the value of V.

By substituting that V value to equation (1) we can find value of f_1. And after that if we substitute that V value, to equation (2) and we can find value of f_2. Then we get the difference of f_1 and f_2. If $f_1 < f_2$; then we can conclude that that asteroid gradually becomes energize(may not be happened) , if $f_2 < f_1$; then we can conclude that the asteroid gradually becomes weaker in energy.

If we assume $f_1 = f_2$ directly, then by substituting that V value to both equations and by comparing those k(V) value with the integer orders of V value we got, we can get rough idea about the function k(V).

Then by using some equations in relative motions and etc. we can estimate how far the asteroid from the Earth roughly.

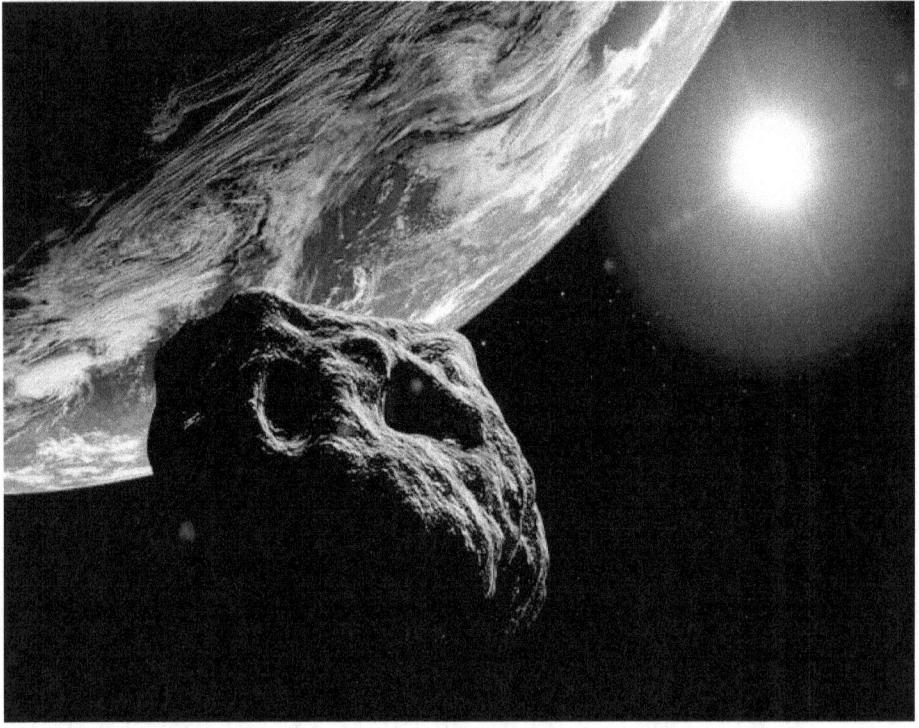

Photo origin: io9.com

Chapter 05

There can be some other force kinds that is stronger than the strong nuclear force, within the nucleus

We know that protons and neutrons within the nucleus are under the influence of strong nuclear force also. Although there are many positively charged protons those have concentrated into a small volume, strong nuclear force can bound them under the power of strong nuclear force.

But, now I'm trying to investigate a new idea regarding the strong nuclear force and deep consideration rather than strong nuclear force. We know that with the development of science, scientists could figure out small particles in nature smaller than previously found. Then we can hope there would be an investigation relate to such a small particle rather than previous. Let us assume scientists could find such a particle (say 'X') smaller than present findings in the year 2020. We had the smallest building particle of the nature (according to our knowledge) as 'Y' in the year 2013. We know that strong nuclear force is the present strongest force among the four fundamental forces in nature. That acts on smallest particle in our nature. Let us assume we would have to consider the strong nuclear force as the strongest force in our nature in the year 2020. Then we could find X in the year 2013 also through nuclear fission procedures. However, we have assumed that in 2013, the smallest building particle our nature is Y. Then it is a contradiction. That clearly implies that there is an another force that is existing within our environment.

But when we consider the whole universe the case would more complicated.

Now I'm trying to come to another aspect rather than the previous. We can get knowledge about new extraordinary fact that is," Earth is not revolving around the Sun". I also believe that fact. Because when the sun is also revolving around the center of the Milky Way galaxy that incident should cause a spiral movement of the Earth. However, if we consider the universal expansion according to the Big Bang Theory, there should be a movement of Milky Way galaxy also through the space-time. That would definitely bring a complexity to the motion procedure of the Earth. This extraordinary fact describes that: 'There should be more number of dimensions within our universe'. Because when we come gradually from the movements of galaxies to the movements of smallest possible building matter in our universe, there should be more complex types of motions within our universe, according to the facts that I mentioned previously (That means according to the notion related to the complex motion of the Earth). There can be more dimensions in our universe. String theory explains a new extraordinary phenomenon that reveals the importance of multi-dimensional world. In addition, it deals with Big Band theory as well as with the procedures those are carrying out the pre-behaviors of the space in the early universe.

String theory emphasizes a new notion regarding some features of fundamental particles in nature. It investigates all the particles in nature have build-up with same basis (call strings) and also depending on the vibration patterns of those basis, fundamental matter types vary. Also reveals: more dimensions have existed in early universe. However, as the attempt of time flowing, those extra dimensions in space have curved up to some critical size. Now a days we feel only less number of dimensions indeed. According to my previous own argument, development of science may express how to adapt the string theory, to carry out the usage of multi-dimensional world.

Multi dimensions in String Theory (Photo credit:
http://www.pbs.org/wgbh/nova/physics/imagining-other-dimensions.html

Chapter 06

Ways to apply the radiations get from the radioactive decays for day to day activities and experiments

Indeed when comparing ancient era and recent time, science has not changed. But, the observable change between those eras caused by the human brain evolution and with the new researches. That means although the science is entirely same actually, the feeling of change of science brought by human mind. That means science is just like 'find out gems from a sand land'.

With a development of science, the damages those are possible to influence for humankind also raised. As examples, radiation damages cause on human body, nuclear power radiation damages, electric power damages cause on human body are the major ones. But, for some aspects of such damages, scientist's attempts to find solutions have become possible indeed.

The nuclear power damage is one of the most dangerous aspect. Because, nuclear power sources have very strong harmful power to damage human bodies. When people are testing those kinds of experiments such as nuclear radiation power testing experiments, the probability of harmfulness is a very high value. Extraordinary, at the airports, governments usually use the nuclear decay concept and also use the x ray radiation testers in order to give the legal certificate when people are traveling from country to country. Governments ensure the airport services for those testing, due to the less harmfulness. Now a days nuclear decay applications play important role in market sales, airports, when goods import to a country, testing the quality of medicines before selling and etc.

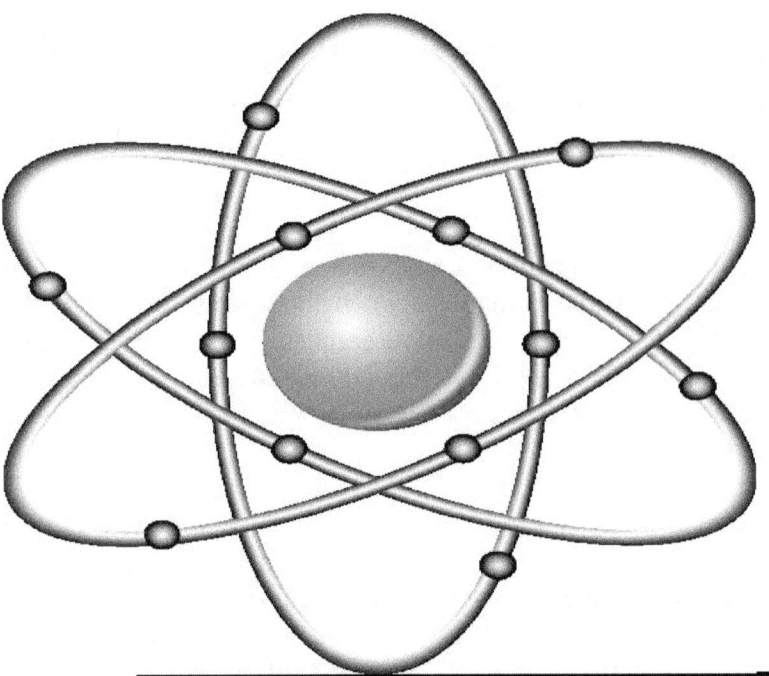

Photo credit:http://www.aip.org/history/exhibits/rutherford/sections/alpha-particles-atom.html

Photo credit: www.dreamstime.com (**Nuclear energy creating**)

Some elements (such as Th, Po, U and etc.) in periodic table have a special property called the radioactive property. Radioactive elements in periodic table are usually very unstable under usual circumferences. (Such as temperature, pressure). Due to that, they radically decay to another elements those are physically stable in the periodic table. When the decay process is active for such radioactive elements; high energy photon such as ϒ (Gamma) ray or a neutrino particle should emit in order to conserve the mass- energy conservation law in nature. For nuclear decay processes, those emitting photons have very high radiation power in order to resolve and in order to damage the human cells. Basically, nuclear radiations help to create cancer cells within human body due to the high energy as well as high resolution power. And, if those high energy ϒ ray photons incident on human eye, those can influence to the human eye lenses also. Because we know that human eye lenses can entirely different from the lenses those have made in laboratories. Human eye lenses can change the focal length according to the distance to the observing object. Basically those nuclear radiations can change that property of human eyes easily. And also nuclear radioactive effects have lot of advantages for mankind.

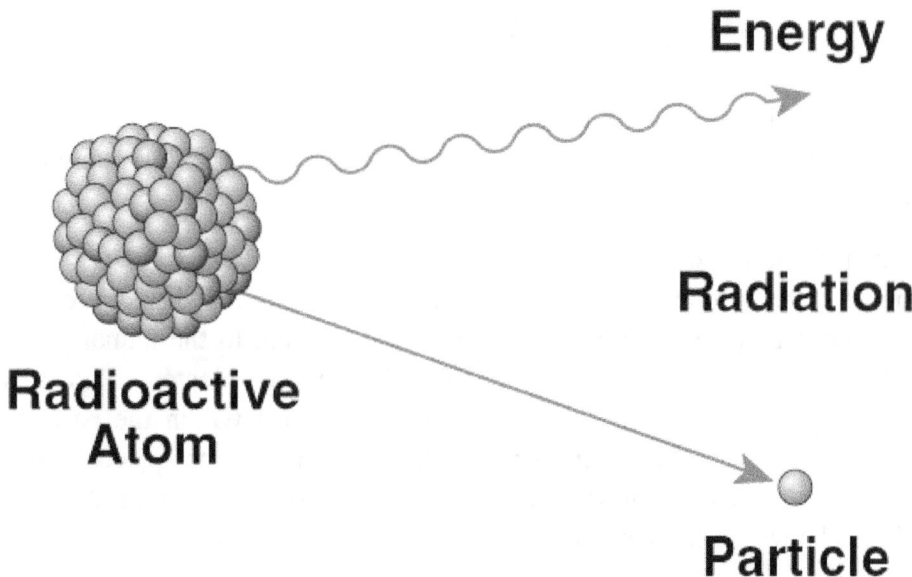

Nuclear decay basics (Photo credit: http://www.nrc.gov/reading-rm/basic-ref/glossary/full-text.html)

If some food samples have affected by the harmful radioactive elements, then such food samples can easily detect. Because from those food samples, we can detect high energy electromagnetic photons or high energy particles (ex: neutrinos). If we do those experiment within a laboratory, we can use a ZnS layered plate. When such a high energy particles incident on the ZnS layered plate, we can observe that easily.

ZnS layered plate. Photo credit: http://www.what-is
nanotechnology.com/ZnS+Ag+Mn+ZrO2+ZnO+Gd2O3+Y2O3+targets.ht
m

And, radioactive decay processes apply for medical purposes also. If a doctor wants to check a place within a human body that has any cancer cells, doctor injects less harmful radioactive element to the human body. But after some time, (let after a half- life of the injected radioactive element), doctor can check whether particular place within the patient's body emits the nuclear radiations rapidly rather than other places within the patient's body or not. Basically, cancer cells are very active rather than other cells within human body.

Then cancer cells are active more for the radioactive isotopes rather than other cells within human body. Then radioactive elements rapidly decay within those cancer cells rather than other cells in the human body.

And, also I think according to my own thoughts , people can use radioactive property of some elements in the periodic table, in order to check whether ; within some food or some material sample are there any less useful elements included or not. Because we know that, after the half-life has gone of those radioactive elements those are in those testing sample, depending on the material of the food or depending on the material of the good sample, the intensity of detecting high energy radiations or intensity of detecting high energy particles can vary. Because we know that, the radiation emitting property and the high energy particle emitting property from a radioactive element, should depend on the neighboring chemical element type (Such as the atomic density of neighboring atoms, atomic radius of neighboring atoms and etc.).

And, there are another very useful aspects of nuclear physics. But those are usually do not depend on the nuclear radioactive property of the radioactive elements. Those are nuclear fusion and nuclear fission. In the recent time, the nuclear fission is a procedure of generating large amount of power for the usages of power. But basically the nuclear fusion procedure apply as the energy generating procedure inside the core of a star and within the space objects those have high mass and high temperature inside(As energy self – generating procedure).

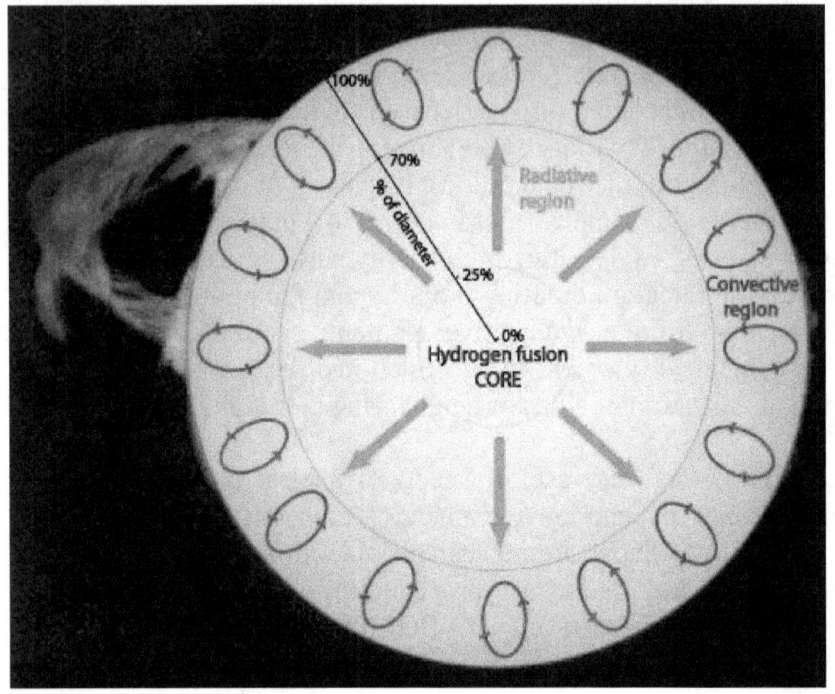

Main energy generating sequence of a star
Photo Credit: Background image of the Sun at 30.4 nm in extreme ultraviolet taken from Skylab in 1973 provided courtesy of the Naval Research Laboratory

Chapter 07

Conclusions of General theory of relativity

Abstract

According to General theory of Relativity, near a high massive object, the space-time curves under the influence of high gravitational field of the mass of the high massive object. Also, under the influence of curved space-time, the light rays those are moving near to the high mass should bend.

According to the general theory of relativity, whenever a star has attained the specific radius `Schwarzschild Radius', the light rays those are going outward from the star would not receive by the external observer. That named as 'infinite gravitational red-shift'. If the light ray emitting star (That has attained `Schwarzschild Radius') is moving with some non-zero velocity relative to the Observer, then there is a possibility to observe the 'finite gravitational red shift' according to the conclusions of this chapter.

Since all the objects are moving away from each other, we can consider that any light ray emitting star has none-zero velocity relative to other objects in the universe. Therefore the practical applications of my argument are not very rare.

7.1. Introduction

7.1.1.Background

Science is a logical and analytical subject field that deals with two aspects. The past explorations and the recent researches and explorations. With the development of science, scientists reveal much complex and much valid concepts and theories those can challenge to the classical explorations. Ancient human dealt with several fundamental un-known concepts and theories although they did not have a clear idea regarding the principles those are related to the field.

Newton who discovered the notion 'Gravity', argued with the thing he saw in the environment and analyzed those facts scientifically. Ultimately, he investigated the concepts regarding 'Gravity' and he built up a theory called the 'Newton's theory of gravity'. But, universal law of Newton's gravity theorem consider as a classical theory of gravity. In 1915, great scientist Albert Einstein revealed a theory regarding the features and applications of gravity. The General theory of gravity is different from the Newton's theory of gravity, since Newton's theory of gravity is a classical theory.

Newton's argument was that any matter on the surface of the Earth has influenced by the Earth's gravitational field up to some distance from the surface of the Earth due to gravitational field affects on that matter.

Moreover he introduced a formula that figured out the gravitational attraction between two matter objects. That formula called as Newton's Universal gravitational theorem as below.

$$F = G.M.m/ r^2 \dots\dots\dots\dots\dots\dots 7.1.1.1$$

Where F is the gravitational force between the two objects, G is the universal gravitational constant, M are m are the masses of the two objects, r is the distance between two objects.

Einstein came up with the idea that has related to the space-time curvature. He formulated a relationship between space-time curvature and the mass energy in the universe. His idea was that a large mass can curve the space-time around it. As a result of the space-time curvature around a star, the light rays those are going near to the high massive object should bend and the observer can see an apparent position of the star rather than its true position.

Later, scientists observed several number of images of one single star in the sky. They identified that this is due to the gravitational light bending. They called this incident as 'Gravitational lensing'

Einstein's famous formula in general theory of relativity as below.

$$R_{ab} - (1/2)Rg_{ab} = 8\pi.GT_{ab}/C^4 \dots\dots\dots\dots\dots 7.1.1.2$$

This formula is valid whenever the cosmological constant is zero.

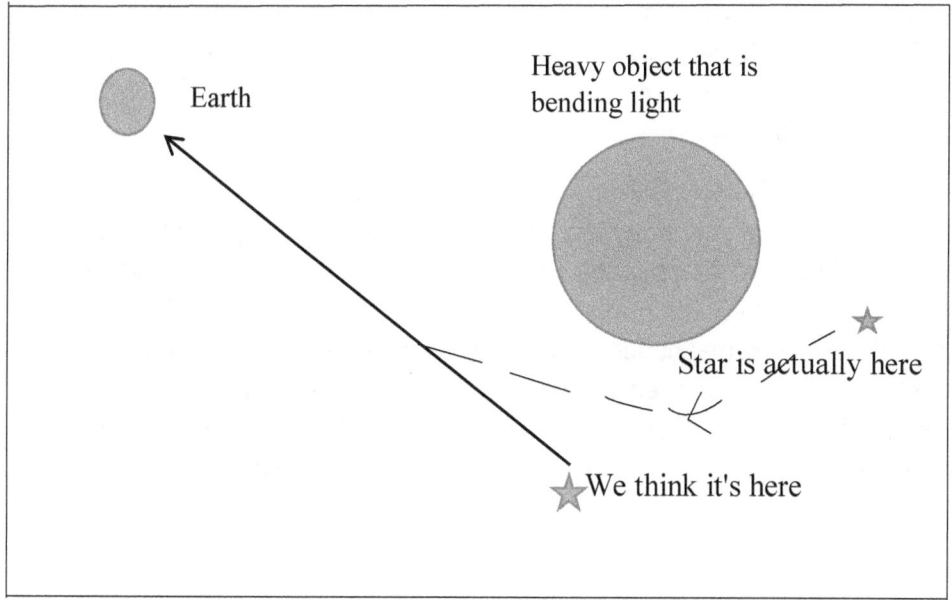

Figure 7.1. Gravitational light bending near a high massive star

7.1.2. Literature survey of computer simulations and relation to the study

Scientists have designed several kinds of computer simulations and soft wares those consist of relativity's notions by using numerical methods, algorithms to solve and analyze. They used supercomputers to study the black-holes, neutron stars, Pulsars, gravitational waves and etc.

The main goal of numerical relativity is to study the un-known space-time. The space time has designed computationally as fully dynamical or stationary or static. When we consider stationary and static solutions, numerical methods are so useful to study the stability of the equilibrium of space times. The dynamical space-times may be divided into the initial value problem and the evolutions.

Numerical relativity is so important in many areas, such as cosmological models, perturbed black holes and neutron stars, and the coalescence of black holes and highly dense neutron stars. In those cases, Einstein's equations can convert and formulate in some ways.

In the problems of numerical analysis, we consider the stability and convergence of the numerical solutions. In this case, the attention should paid to the coordinates and various formulations of the Einstein equations and the effects those have the capability to come up with accurate numerical solutions.

Numerical relativity research is different from the work of classical field theories. Theories and many techniques and related concepts, implemented in these areas are inapplicable in relativity. There are several activities those shared with computational sciences like computational fluid dynamics, electromagnetics and solid mechanics. Numerical relativists usually work with applied mathematicians in order to draw insight from numerical analysis, scientific computation, partial differential equations and geometry among other mathematical areas of specialization.

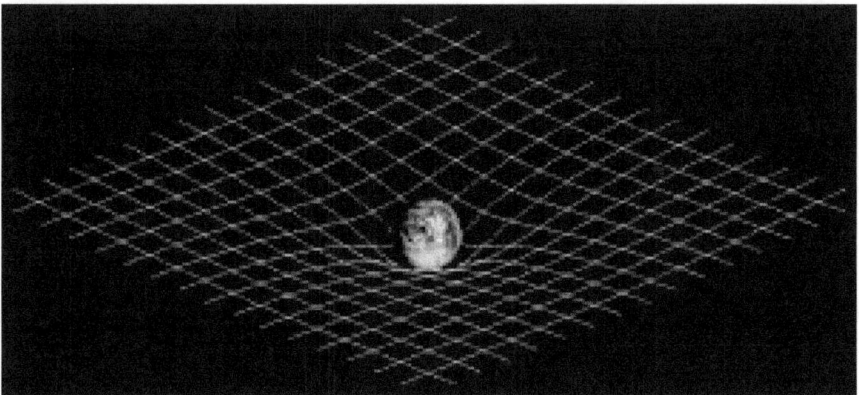

Figure 7.2. Computer simulations in general theory of relativity Photo origin: https://en.wikipedia.org/wiki/Gravitational_singularity

The study of astrophysical phenomena can involve with several areas of general theory of relativity. Experimental astrophysics and cosmology researches can involve with several areas of experimental physics, such as designing, building, and observing with new telescopes and cryogenic receivers, developing and testing superconducting devices and analyzing data from observatories to help address questions about our galaxy and the universe. Since General theory of relativity has involved in all the areas of astrophysics all most, the applications of the theory are useful for all above.

Discussing for researching and challenging the principle of general covariance are so worthy (which has started purely from mathematical perfection, in 1974 Mao Zedong indicated, he completely can't understand that symmetry can hold in both hands to the so high status in physics). In fact, the symmetrical theory relies on the extremely profound mathematical tool, purely for the purpose of popularization of the asymmetrical theory in the aspect of mathematical perfection that must be developed (but it conforms to the symmetry in the physical aspect, such as the principle of conservation of energy and so on). Starting from the physical significance instead of mathematical perfection, one can develop and apply this kind of asymmetrical theory to substitute for the general theory of relativity and also other symmetrical theory. This is possibly of a correct direction indeed. Some students already stride out the important and solid steps to establish this kind of asymmetrical theory to substitute for the theory of relativity. Moreover, in views of the principle of equivalence, the space-time theory, the principle of relativity, the principle of invariance of light speed and the ultra-speed of light, the Lorentz transformation equations, the certain results of relativity (such as ruler shrinking and clock slowing and so on) as well as question of unified theory, elaborates the main achievements for researching and challenging the general theory of relativity.

The present researches in the gravitational light bending

At the beginning of 1915, the general theory of relativity did not have a solid empirical foundation indeed. It was known that it correctly accounted for the "anomalous" precession of perihelion of Mercury and on philosophical grounds, was considered to unify Newton's theory of gravity (which is a classical theory) with special theory of relativity. That considered light appeared to bend in gravitational fields in line with the predictions of general theory that was found in 1919. But it was not a final program of precision tests that started in 1959. But the various types of predictions of general theory of relativity were tested to any further degree of accuracy in the weak gravitational field limit, severely limiting possible deviations from the theory.

The very strong gravitational fields (such as black holes, especially those super-massive black holes which are thought as power active galactic nuclei and the more active quasars), belong to a field of intense of active research. Observations of these quasars and active galactic nuclei are difficult indeed. And interpretation of the experimental observations are heavily dependent upon astrophysical models other than general theory of relativity. But they are quantitatively consistent with the black hole concept, event horizon as modeled in general relativity.

Especially, as a consequence of the equivalence principle, Lorentz invariance holds locally in freely falling reference frames. Experiments related to Lorentz invariance and special relativity are described in tests of special relativity.

Experimental verification of the gravitational red-shift requires good clocks since at Earth the effect is small. The first experimental confirmation has done in 1960. The famous experiment is generally called the Pound-Rebka experiment. Through that, they used a well-defined "clock" in the form of an atomic transition which results in a very narrow line of electromagnetic radiation. A narrow line implies a very well defined frequency. The line is in gamma ray range and emitted from the isotope Fe57 at 14.4 keV. The narrowness of the line is caused by the so called Mossbauer effect.

7.1.3. Gravitational light bending

The definition

Space time around a massive object such as a galaxy cluster, near a neutron star or a black hole is curved and as a result, the light rays propagating near to the massive object (such as a galaxy) bend. The 'lensing effect' occurs due to the gravitational light bending.

As a result of gravitational light bending, an observer can see several images of one light emitting star. That incident called as 'gravitational lensing'.

Unlike optical lenses, maximum bending occurs closet to the center of a gravitational lenses. Moreover, gravitational lenses have no single focal point, but focal lines are there. If the light source, the massive lensing object, and the observer lie in a straight line, the original light source will appear as a ring around the massive lensing object.

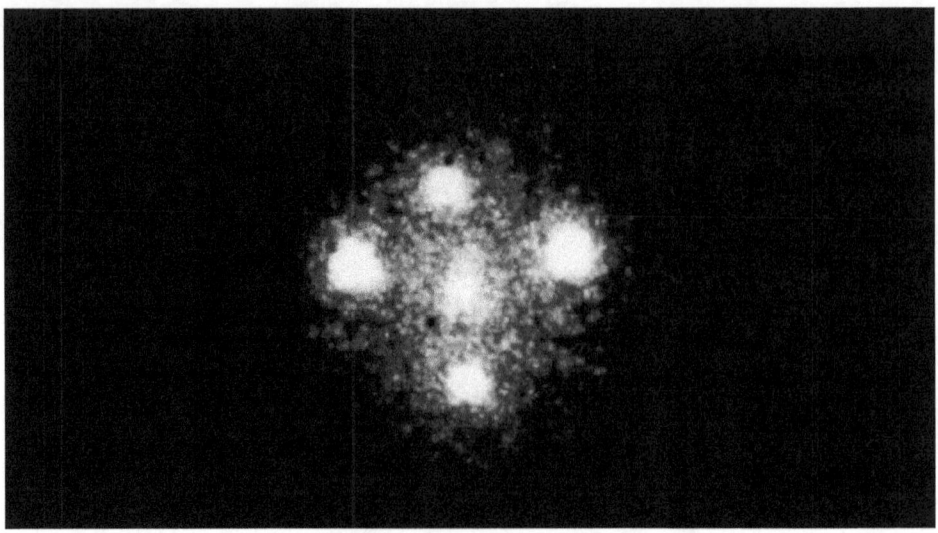

Figure 7.3. Gravitational lensing

Photo origin: https://en.wikipedia.org/wiki/Gravitational_lens

Detecting the gravitational light bending

In order to detect gravitational light bending properly (The gravitational light bending due to the mass of the sun), observers should wait until a full solar eclipse comes. Because other than solar eclipse, scientists can't detect the path of the light ray that is coming from a distant star. Therefore in order to verify Einstein's General theory of relativity, scientists had to wait until a full solar eclipse comes. In 1919, when the full solar eclipse was coming, scientists were able to verify Einstein's General theory of relativity. After that also, some scientists have verified Einstein's General theory of relativity more accurately. Later, scientists did able to detect gravitational lenses those are due to the gravitational light bending.

Major part of the gravitational lenses in the past, were discovered by doing experiments accidently. A research for gravitational lenses in the northern hemisphere done through Radio frequencies using the Very Large Array(VLA) in new Mexico, led to the discovery of new lensing systems, a major milestone. This is an extraordinary experiment for the researches, which is from finding very distant objects to finding values for cosmological parameters. So we can understand the universal concepts and theories better.

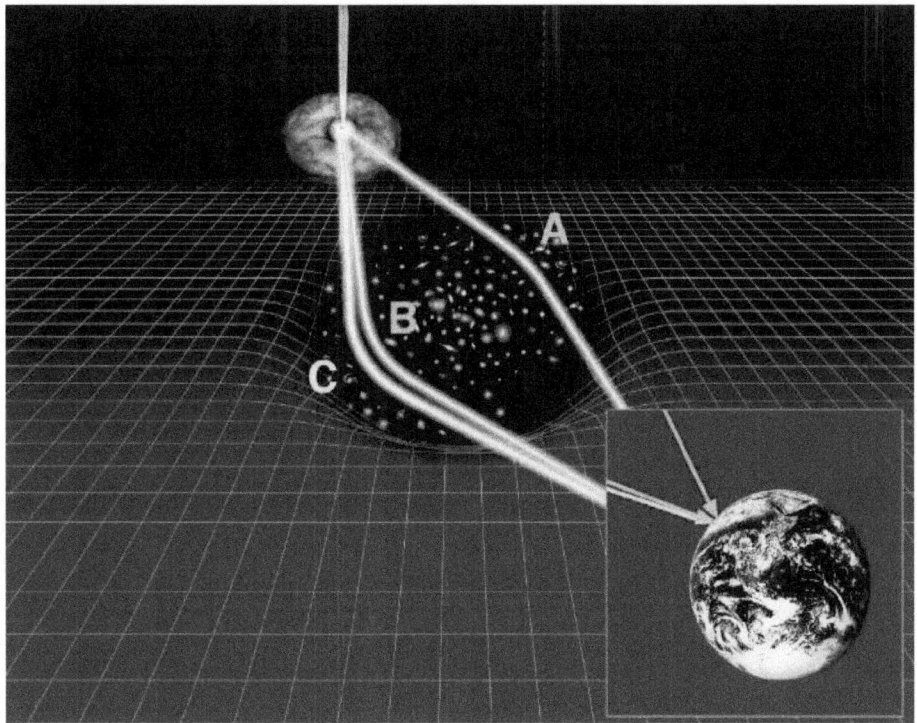

Figure 7.4. Verification of gravitational light bending

Photo credit:

http://subarutelescope.org/Pressrelease/2013/02/18/index.html

The applications of gravitational light bending

In general relativity, the presence of matter (energy density) can curve space-time, and the path of a light ray will be deflected as a result. This process called as gravitational lensing and in many cases it can be described in analogy to the deflection of light by lenses in optics. Many useful results for Cosmology came to the world using this property of matter and light basically.

For many of the cases of interest, one does not need to fully solve the general relativistic equations of motion for the coupled space-time and matter. Because the bending of space-time by matter is small.

Einstein investigated the general theory of relativity in 1919. But after that scientists could understand the various applications of it. Researches have fully covered Gravitational light bending through the general theory of relativity. Whenever the scientists do the observations of gravitational light bending concepts, gravitational wave phenomenon they use as the basis. To explain several unusual images of distant stars as well as for several space explorations and space missions, the theory is essential. Whenever the scientists proceed with deep space explorations, they use the methods in general relativity in order to focus their space shuttle to the relevant position in space. The decrease of the orbital period of some planets those are orbiting around a star, cannot explain using classical theories. Because that extraordinary incident has related to the gravitational wave phenomenon. Whenever a planet is orbiting around the star, it disturbs to the space time around its orbit. Then gravitational waves emit. Due to the loosing energy of the planet, the orbital period gradually decreases. That incident has an explanatory related to the general theory of relativity.

7.2. Theory

7.2.1. Gravity

7.2.1.1. Definition

Gravitation or gravity is a natural phenomenon by which all physical bodies attract each other. It is most commonly experienced as the agent that gives weight to objects with mass and causes them to fall to the ground when dropped.

Gravitation is one of the four fundamental interactions of nature, along with electromagnetism and the nuclear strong force and weak force. In modern physics, the phenomenon of gravitation has most accurately described by the general theory of relativity of Einstein, in which the phenomenon itself is a consequence of the curvature of space-time governing the motion of inertial objects in the universe. The simpler Newton's law of universal gravitation postulates that the gravity force proportional to masses of interacting bodies and inversely proportional to the square of the distance between them. It provides an accurate approximation for most physical situations including calculations as critical as spacecraft trajectory.

From a cosmological perspective, gravitation causes dispersed matter to coalesce and coalesced matter to remain intact, thus accounting for the existence of planets, stars, galaxies, nebulas and most of the macroscopic objects in the universe. It is responsible for keeping the Earth and the other planets in their orbits around the sun; for keeping the Moon in its orbit around the Earth; for the formulation of tides and various other phenomenon and observations observed on Earth and throughout the universe.

7.2.1.2. Light bending

As described by Einstein, high massive object can curve space-time around it. Therefore whenever a light ray comes near to that high massive object; it has been influenced by the curved space-time near the high massive object. Thus the path of the light ray bends towards the high massive object. Therefore an observer able to see an apparent position of the star rather than its true position. This special incident calls as gravitational light bending.

7.2.1.3. Infinite gravitational red shift

When a star is emitting light rays towards an observer, usually those light rays will reach the observer at some time in future. If the star emits light rays with a time duration Δt, usually those light rays approach to the observer with the same time duration. Therefore usually the frequency of those receiving light rays by the observer are not changed and same as the frequencies of the emitted ones. But this not happens if the light emitting star has attained some specific radius that is a critical value. Then the space-time around it gets curved. Therefore the time duration between the consecutive receiving light rays by the observer gradually increases with the decrease of the radius of the star. Therefore the frequencies of receiving light rays by the observer gradually decreases. Therefore the wavelength's of those receiving light rays gradually increases. This is due to the highly curved space-time around the light emitting star. Also, if the light emitting star has attained some specific radius calls `Schwarzschild radius'(This specific radius limit calls as `event horizon' also), the space-time around it gets highly curved. Therefore the time duration between the consecutive receiving light rays those are receiving by the observer tends to infinity. Therefore the frequency of receiving light rays by the observer tends to zero. Therefore the light rays those are emitting by the Schwarzschild radius attained star, would not receive by the observer ever. Therefore the wavelength's of those receiving light rays have been highly red-shifted. This is due to the high gravitational force of the star. Therefore this incident calls as infinite gravitational red shift.

7.2.1.4. The features and origin of gravity

Nowadays the most believing theory of the origin of the universe is Big Bang theory. According to Big Bang theory, all the matters in the universe came to exist, after the Big Bang explosion happened. Therefore, we have to consider the origin of gravity basically as the Big Bang explosion.

It is already proven fact that the speed of light is reduced in a gravitational field. As a consequence, a light beam which passes a massive object bends towards that object. This specific deflection can be explained quantitatively by the refraction of the light rays due to the gravitational potential.

From the spin and the magnetic moment of an elementary particle, it is clear that the constituents of such particles are constantly oscillating while traveling at the speed of light according to the recent discoveries. If we consider an electron, this has researched by Paul Dirac. And the effect of refraction is applied to this oscillation within an elementary particle, and it would give the correct gravitational acceleration within a gravitational field area.

This calculation not only directs the gravitational behavior of an object which is at rest but it explains the acceleration of fast moving objects in a gravitational field area. Which has usually explained by using the Einstein's general theory of relativity. The refraction of light-like objects by the field is an equivalence. Moreover, very surprisingly, scientists have found that the mass of an object is not the cause of its gravitational field according to General theory o relativity.

This consequence also motivated the scientists to do researches in dark matter and dark energy.

7.2.1.5. Where the applications of gravity occur

The applications of gravity occur everywhere the matters have located in the universe. But, the interaction region of the gravitational force carrier 'Graviton' is almost the whole universe. But near a black hole or high massive star, the gravitational field strength is high. But in the micro-gravity environment, the gravitational field strength is somewhat low. Usually, we observe micro-gravity environment beyond the Earth's surface and in the environments just 30000 km above from the surface of the Earth.

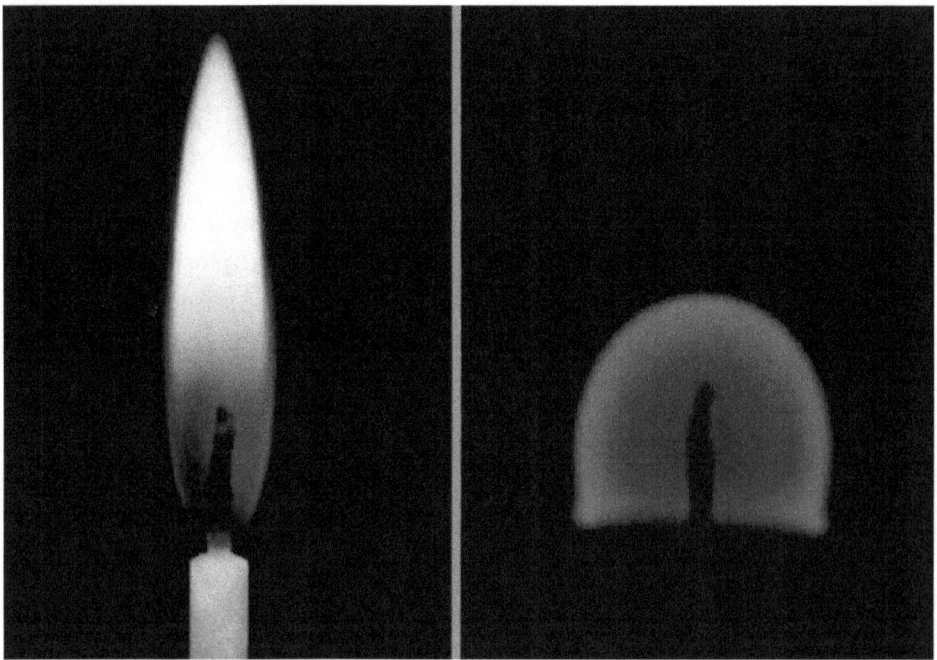

Figure 7.2.1. Microgravity environment

Photo origin: https://commons.wikimedia.org/wiki/File:Space_Fire.jpg

Figure 7.2.2. Microgravity environment

(Photo origin: http://genelab.nasa.gov/about.html**)**

7.2.2.1. Description to General theory of relativity

In 1915 Albert Einstein introduced remarkable theory called as 'General theory of relativity' as a research article. This theory was a challengable theory regarding gravity as well as for other classical theories in Physics.

Before the advent of general relativity, Newton's law of universal gravitation had been accepted for more than two hundred years as a valid description of the gravitational force between masses, even though Newton himself did not regard the theory as the final word on the nature of gravity. Within a century of Newton's formulation, careful astronomical observation revealed unexplainable variations between the theory and the observations accurately. Under Newton's model, gravity was the result of an attractive force between massive objects.

However, experiments and observations show that Einstein's description accounts for several effects those are unexplained by Newton's law. General relativity also predicts novel effects of gravity, such as gravitational waves, gravitational lensing and an effect of gravity on time known as gravitational time dilation. Many of these predictions have been confirmed experimentally. For example, although there is an indirect evidence for gravitational waves, direct evidence of their existence is still being sought by several teams of scientists through experiments.

General theory of relativity has developed into an essential tool in modern astrophysics. It provides the foundation for the current understanding of black holes, regions of space where gravitational attraction is so strong that not even light can escape. Their strong gravity is thought to be responsible for the intense radiation emitted by certain types of astronomical objects (Such as active galactic nuclei or micro quasars). General relativity is also a part of the framework of the standard Big Bang model of cosmology.

7.2.2.2. Equations in General Theory of relativity

Einstein built General Theory of Relativity by using several mathematical equations and methods such as Tensors and Metrics.

The Einstein filed equations (EFE) are set of 10 equations which describe the fundamental interaction of gravitation as a result of space-time that being curved by matter and energy. It first published by Einstein in 1915 as a tensor equation. This remarkable EFE equates local space-time curvature with the local energy and momentum within that space-time (expressed by the stress-energy tensor). Similar to the way that the electromagnetic fields determined using charges and currents via Maxwell's equations, the EFE are used to determine the space-time geometry resulting from the presence of mass-energy and linear momentum in the universe. The relationship between the metric tensor and the Einstein tensor allows the EFE to be written as a set of non-linear partial differential equations. The solutions of the EFE are the components of the metric tensor as described in general theory of relativity.

As well as obeying local energy-momentum conservation, the EFE can reduces to Newton's law of gravitation where the gravitational field is weak and velocities are much less than the speed of light. Exact solutions for the EFE can only be found under simplifying assumptions such as symmetry. Special classes of exact solutions have studied as they model many gravitational phenomena, such as rotating black holes and the expanding universe as indicated through recent researches. Further simplification is achieved by approximating the actual space-time as flat space-time with a small deviation, leading to the linearized EFE. These equations are used to study phenomena such as gravitational waves.

7.2.2.3. Mathematical tool of general relativity

Prof. Albert Einstein (Photo credit:
http://www.nobelprize.org/nobel_prizes/physics/laureates/1921)

Indices are more convenient ways of writing the components of a vector. Consider the following vector.

$$r = r^0e_0 + r^1e_1 + r^2e_2 \quad \dots\dots\dots\dots\dots\dots\dots 7.2.1$$

Which has three components. These components can be written in a simpler form using indices,

$$r = r^\mu e_\mu \quad \dots\dots\dots\dots\dots\dots\dots\dots\dots\dots 7.2.2$$

To show that Equation (7.2.2) is equivalent to Equation (7.2.1), we introduce the summation convention, which states that: repeated upper and lower indices are added over in any given expression. Using this definition we obtain the following representation of equation 7.2.2

$$r = \sum_{\mu=0}^{\mu=2} r\mu e\mu\dots\dots\dots\dots\dots\dots\dots\dots\dots\dots\dots.7.2.3$$

All of the work done with Einstein's Field Equation starts with analyzation of metrics*[see appendix]. A metric generally provides informations required to calculate distance between two points, which is given by ds^2, defined as:

$$ds^2 = g_{\mu v}\, dx_\mu dx^v\dots\dots\dots\dots\dots\dots\dots\dots..7.2.4$$

*** Please refer the meanings of those tensors $g_{\mu v}$, $R^\alpha{}_{\mu\alpha v}$, $\Gamma^\alpha{}_{\mu v,\alpha}$ and other tensors in General theory of relativity.

In the Equation (7.2.4), the two indices μ and v represent the components of the metric. Writing metrics in matrix form make the concept of two different indices clearer. The structure of the metric $g_{\mu v}$ depends on the physical quantity that we are considering. So, for example, if we are talking about flat space-time, these indices will be dependent upon the components x,y,z and t (time). We can write the metric in matrix form by making use of the summation convention to obtain the matrix:

$$g_{\mu v} = \begin{bmatrix} g_{tt} & g_{tx} & g_{ty} & g_{tz} \\ g_{xt} & g_{xx} & g_{xy} & g_{xz} \\ g_{yt} & g_{yx} & g_{yy} & g_{yz} \\ g_{zt} & g_{zx} & g_{zy} & g_{zz} \end{bmatrix} \dots\dots\dots\dots\dots\dots.7.2.5$$

We are now in a position to provide an example of a space-time metric: flat space-time is represented by the Minkowski Metric. The Minkowski Metric has four components x, y, z and time t:

$$ds^2 = -(c.dt)^2 + dx^2 + dy^2 + dz^2\dots\dots\dots\dots\dots\dots.7.2.6$$

Looking at the setup in Equation (7.2.5), we can write this metric in matrix form as:

$$g_{\mu v} = \begin{bmatrix} -c^{-1} & 0 & 0 & 0 \\ 0 & 1 & 0 & 0 \\ 0 & 0 & 1 & 0 \\ 0 & 0 & 0 & 1 \end{bmatrix} \quad \dots\dots\dots\dots\dots\dots 7.2.7$$

Some tensors*[see appendix] may also be written in matrix form as very much like metrics. So, if a tensor $J_{\alpha\beta}$ has the vector components t, r, θ and φ, it could be written in matrix form using the summation convention.

Einstein curvature tensor that describes this curvature in space-time: (the tensor is given by)

$$G_{\mu v} = R_{\mu v} - (1/2)g_{\mu v}.R \dots\dots\dots\dots\dots\dots\dots 7.2.8$$

There exist three components those go from the Einstein curvature tensor: $R_{\mu v}$,$g_{\mu v}$ and R. *** [please refer the meanings of those tensors in General theory of relativity] Out of the three, the more straight-forward one to explain is $g_{\mu v}$. The structure of the metric that we are considering: the Einstein curvature tensor, $G_{\mu v}$, describes the curvature in space-time created by the object represented by the metric, $g_{\mu v}$.

The Ricci curvature tensor, $R_{\mu v}$, roughly speaking, describes the volume of the object in the consideration relative to the number of dimensions of a manifold. The Ricci curvature tensor is a contracted version of the Riemann curvature tensor. Mathematically, the contraction is defined as:

$$R_{\mu\nu} = R^{\alpha}{}_{\mu\alpha\nu} \quad \dots\dots\dots\dots\dots\dots\dots\dots\dots\dots\dots 7.2.9$$

Where $R^{\alpha}{}_{\mu\alpha\nu}$ is the Riemann curvature tensor. The Riemann curvature tensor is given by,

$$R^{\alpha}{}_{\mu\alpha\nu} = \Gamma^{\alpha}{}_{\mu\nu,\alpha} - \Gamma^{\alpha}{}_{\mu\alpha,\nu} + \Gamma^{\alpha}{}_{\beta\alpha}\Gamma^{\beta}{}_{\mu\nu} - \Gamma^{\alpha}{}_{\beta\nu}\Gamma^{\beta}{}_{\mu\alpha}\dots\dots\dots\dots\dots 7.2.10$$

Using the definition of the Riemann curvature tensor, we may apply the summation convention to contract the Riemann tensor to form the Ricci tensor. A contraction involves summing over the repeated upper and lower indices:

$$R_{\mu\nu} = R^{\alpha}{}_{\mu\alpha\nu} = R^{0}{}_{\mu0\nu} + R^{1}{}_{\mu1\nu} + R^{2}{}_{\mu2\nu} + R^{3}{}_{\mu3\nu} \quad \dots\dots\dots\dots\dots 7.2.11$$

The last component of the Einstein curvature tensor is the Ricci scalar (or scalar curvature) given by R. The scalar curvature is a scalar. It represents the curvature of the geometry that we are considering by a single real number. This scalar curvature is given by the following mathematical definition:

$$R = g_{\mu\nu}R_{\mu\nu}\dots\dots\dots\dots\dots\dots\dots\dots\dots 7.2.12$$

Using the summation convention, sum over the indices μ and ν:

$$R = g_{00}R_{00} + g_{01}R_{01} + g_{02}R_{02} + \dots\dots\dots g_{13}R_{13} + g_{32}R_{32}\dots\dots\dots 7.2.13$$

This summation will result in a scalar, which will be a real number.

We take the metric of the geometry that we are considering, find its Christoffel symbols, compute the Riemann curvature tensor, and through that tensor we find both the Ricci curvature tensor and the Ricci scalar. These all the components gives the Einstein curvature tensor:

$G_{\mu\nu} = R_{\mu\nu} - (1/2)g_{\mu\nu}R$

For convenience, let us reproduce the expression or finding the Christoffel Symbols:

$$g_{\alpha\delta}\Gamma^{\delta}{}_{\beta\alpha} = (1/2)[\partial g_{\alpha\beta}/\partial x^{\gamma}] + \partial g_{\alpha\gamma}/\partial x^{\beta} - \partial g_{\beta\gamma}/\partial x^{\alpha}..................7.2.14$$

Where

$$g_{\alpha\delta}\Gamma^{\delta}{}_{\beta\alpha} = g_{\alpha0}\,\Gamma^{0}{}_{\beta\alpha} + g_{\alpha1}\,\Gamma^{1}{}_{\beta\alpha} + g_{\alpha k}\,\Gamma^{k}{}_{\beta\alpha} ....7.2.15$$

$g_{\alpha i}$, can be found by the matrix 7.2.5 depending on the situation that we are considering. $i \in \{0,1,.....k\}$. By equation 7.2.14 and by equation 7.2.15, we can find the values for Christoffel symbols.

Where

$$\Gamma^{\delta}{}_{\beta\gamma} = \Gamma^{\delta}{}_{\gamma\beta} ....7.2.16$$

After found the values for $\Gamma^{\delta}{}_{\beta\gamma}$, it is capable to find the values for $R^{\alpha}{}_{\mu\alpha\nu}$ by equation 7.2.10. Then we have $R_{\mu\nu}$ by equation 7.2.9. By equation 7.2.12, we know the value for R. Then by equation 7.2.8, we know the value of Einstein curvature tensor $G_{\mu\nu}$. Also it can be derive through an expression for the Stress-energy tensor ($T_{\mu\nu}$) in a matrix form. The right hand side of the Einstein field equation (EFE) expresses the mass-energy amount of the considering space-time. And the left hand side of the EFE (Einstein curvature tensor) represents the space-time curvature. The Einstein field equation is as below:

$$R_{\mu v} - (1/2)g_{\mu v}R + g_{\mu v}\Lambda = 8\Pi G.T_{\mu v}/c^4 \dots\dots\dots\dots\dots\dots 7.2.17$$

Where Λ is the cosmological constant that is a very small positive value. Basically, Λ represents the average energy density in the universe. Since the universe's expansion is accelerating, the value of Λ should be a positive value.

Photo origin: http://ancientexplorers.com/author/jasonmartell/

7.2.3. Schwarzschild equation

7.2.3.1. Definition of Schwarzschild equation

In Einstein's general theory of relativity, the Schwarzschild solution (or the Schwarzschild vacuum), is a solution to the Einstein field equations which describes the gravitational field outside a spherical mass (on the assumption that the electric charge of the mass, angular momentum of the mass and universal cosmological constant are all zero). As examples, many stars, planets and black holes including Earth and the sun. The solution has been investigated by Karl Schwarzschild, who first published the solution in 1916.

According to Birkhoff's theorem, the Schwarzschild solution is the most general spherically symmetric, vacuum solution of Einstein field equations. A Schwarzschild black hole or static black hole is a black hole that has no charge or angular momentum. A Schwarzschild black hole has a Schwarzschild metric, and can't be distinguished among other Schwarzschild black hole except its mass.

The Schwarzschild black hole is characterized by a surrounding spherical surface, calls as the event horizon, which is located at the Schwarzschild radius. This is usually calls as the radius of a black hole. Any non-rotating and non-charged mass that is similar than its Schwarzschild radius, forms a black hole. It should know that the solution of the Einstein field equations is valid for any mass M.

In Schwarzschild coordinates, the line element for the Schwarzschild metric has the form:

$$c.d\tau^2 = (1- (r_s/r))c^2dt^2 - (1 (r_s/r))^{-1}dr^2 - r^2(d\theta^2 + \sin^2\theta d\varphi^2)\ldots\ldots7.2.18$$

Where τ is the proper time (time measured by a clock moving along the same world line with the test particle), c is the speed of light, t is the time coordinate (measured by a stationary clock located infinitely far from the massive body), r is the radial coordinate (measured as the circumference, divided by 2Π, of a sphere centered around the massive body), θ is the colatitude (angle from north, in units of radians) and the r_s is the Schwarzschild radius of the massive body, a scale factor which is related to its mass M by $r_s = 2GM/c^2$, where G is the gravitational constant.

The analogue of this result in classical Newtonian theory of gravity, corresponds to the gravitational field around a point particle.

The ratio r_s/r is almost always extremely small. For example, the Schwarzschild radius r_s of the Earth is roughly 8.9 millimeters, while the sun which is $3.3* 10^5$ times as massive has Schwarzschild radius of approximately 3.0 km. the ratio only becomes large close to black holes and other ultra-dense objects such as neutron stars.

The Schwarzschild metric is a solution of Einstein' field equations in empty space, meaning that it is valid only outside the gravitating body. That is for spherical body of radius R. And the solution is valid for r >R. To describe the gravitational field both inside and outside the gravitating body, the Schwarzschild solution should matched with some suitable interior solution at r = R.

7.2.3.2 Applications of the Schwarzschild equation

Schwarzschild equation indicates the Schwarzschild radius. Schwarzschild radius is the radius of a star such that the light rays emitting by the star does not receive by an external observer. That means after the light rays emitting star has attained the Schwarzschild radius, the light rays those are emitting by the star are infinitely red shifted. That means although the star is emitting light rays with a finite interval gap of time, the observer' light receiving frequency is zero. In another words, the Schwarzschild radius (sometimes historically referred to as the gravitational radius) is the radius of a sphere such that, if all the mass of an object is compressed within that sphere, the escape speed from the surface of the sphere would equal the speed of light. Example of an object smaller than its Schwarzschild radius is a black hole. Once a stellar remnant collapses below this radius, light cannot escape and the object is no longer visible. It is a characteristic radius associated with every quantity of mass. The Schwarzschild radius was named after the German astronomer Karl Schwarzschild who calculated this exact solution for the remarkable general theory of relativity.

In 1916, Karl Schwarzschild obtained an exact solution to Einstein's field equations for the gravitational field outside a non-rotating, spherically symmetric body. Using the definition $M = Gm/c^2$, the solution contained a term of the form $1/ (2 M\text{-}r)$. Where the value of r is zero, that point has been called as the singularity where the physics laws breaks. The physical significance of this singularity and whether this singularity could ever occur in nature, has described for many decades; a general acceptance of the possibility of a black hole did not occur until the ending half of the 20th century.

7.2.3.3. Formula for Schwarzschild radius

The Schwarzschild radius is proportional to the mass with a proportionality constant involving the gravitational constant and the speed of light.

$$r_s = 2Gm/ c^2 \dots\dots\dots\dots\dots\dots\dots\dots\dots\dots\dots\dots\dots\dots 7.2.19$$

Where r_s is the Schwarzschild radius, G is the gravitational constant, m is the mass of the object, c is the speed of light in vacuum.

The proportionality constant, $2G/c^2$, is approximately 1.48* 10-27 m/kg, or 2.95 km/solar mass.

Derivation of Schwarzschild radius

The derivation of the Schwarzschild radius comes from the fact that at the event horizon the gravitational potential energy is equal to the kinetic energy.

The force of attraction between to massive bodies due to gravity is

$$F = GMm/r^2 \ldots\ldots\ldots\ldots\ldots\ldots\ldots\ldots\ldots\ldots\ldots\ldots\ldots 7.2.20$$

The centripetal force of an object in orbit is $F = mv^2/r$ 7.2.21

Imagining a particle orbiting a singularity at the speed of light

$GMm/r^2 = mv^2/r$ and $v = c$, for the satisfaction of Schwarzschild radius:

$$r_s = 2Gm/c^2 \ldots\ldots\ldots\ldots\ldots\ldots..\ldots\ldots\ldots\ldots\ldots\ldots 7.2.22$$

Schwarzschild radius(m)	Density (gcm^{-3})
Universe $4.46* 10^{25}$	$9.9 * 10^{-30}$
Milky way $2.08 * 10^{15}$	$3.72 *10^{-8}$
Sun $2.95 * 10^3$	$1.84 *10^{16}$
Earth $8.87 * 10^{-3}$	$2.04 * 10^{27}$

Table 7.2.1 Schwarzschild radius

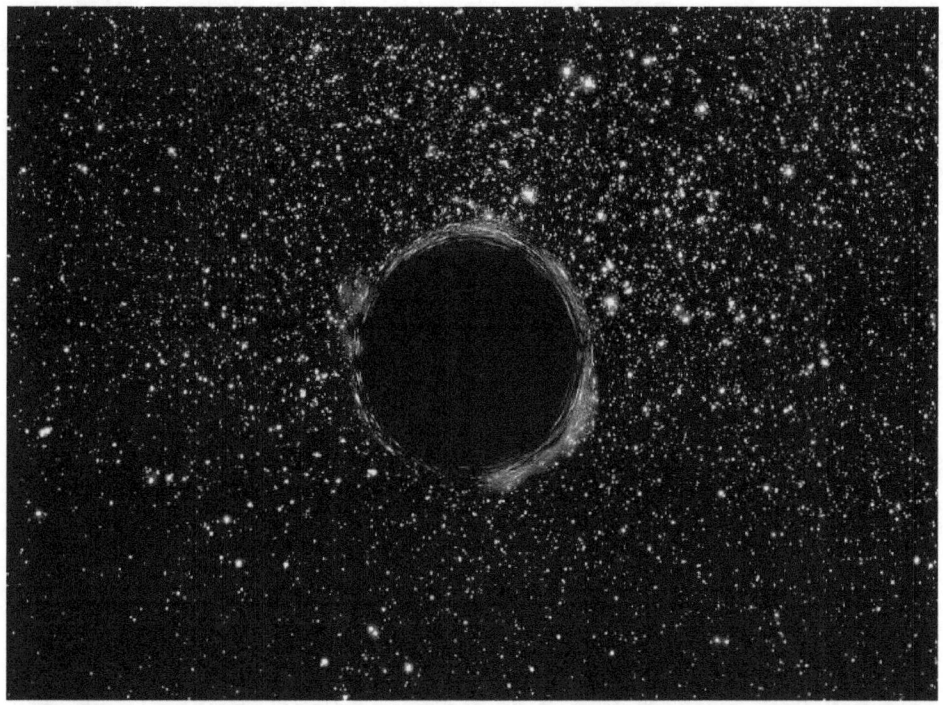

Evidence for a black hole (Photo credit: NASA.gov)

7.2.4. Infinite gravitational red shift

7.2.4.1. Definition of infinite gravitational red shift

Gravitational red-shift or Einstein shift is the process by which electromagnetic radiation emitting from the source (that generates a strong gravitational field) and the observer who is in a region of a weaker gravitational field. This is a direct result of gravitational time- dilation as one moves away from a source of gravitational field, the rate at which time passes is increased relative to the case when one is near to the source. We know that the frequency is the inverse of time (specifically, time required for completing one wave oscillation), and the frequency of the electromagnetic radiation is reduced in an area of a higher gravitational potential (i.e. equivalently of lower gravitational field). There is a corresponding reduction in energy when electromagnetic radiation is red-shifted, according to Plank's relation (due to the electromagnetic radiation propagating in opposition to the gravitational gradient). There also exists a corresponding blue-shift when electromagnetic radiations propagating from an area of a weaker gravitational field to an area of a stronger gravitational field.

When we applied this to optical wavelengths, this manifests itself as a change in the color of visible light as the wavelength of the light is increased toward the red part of the light spectrum. Since frequency and wavelength are inversely proportional, this is equivalent to saying that the frequency of the light has reduced towards the red part of the light spectrum, giving this phenomenon the name red-shift. When the light emitting star has attained the Schwarzschild radius, that gravitational red shift becomes infinite. That gravitational red shift is called as `Infinite Gravitational red shift'.

Red shift is often denoted with the dimensionless variable z, defined as the fractional change of the wave length.

$$z = (\lambda_0 - \lambda_e)/\lambda_e \dots\dots\dots\dots\dots\dots\dots\dots\dots\dots\dots\dots\dots\dots\dots 7.2.23$$

Where λ_0 is the wavelength of the electromagnetic radiation (photon) as measured by the observer. λ_e is the wavelength of the electromagnetic radiation (photon) when measured at the source of emission.

7.2.4.2. The theories behind infinite gravitational red shift

The gravitational red shift of a photon can be calculated in the framework of General Relativity (using Schwarzschild metric) as

$$\lim r \rightarrow +\infinity \quad z(r) = (1/ \sqrt{(1 - r_s/R^*)})$$

With the Schwarzschild radius $r_s = 2GM / c^2$ $\dots\dots\dots\dots 7.2.24$

When $R^* = r_s$, z(r) becomes infinity. That means when the radius of light emitting star has attained Schwarzschild radius, the gravitational red shift becomes infinity.

Where G denotes Newton's gravitational constant, M the mass of the gravitating body, C is the speed of light, and the R^* the distance between the center of mass of the gravitating body and the point at which the photon is emitted. The red shift is not defined for photons those emitted inside the Schwarzschild radius, where the distance from the body where the escape velocity is greater than the speed of light.

Therefore this formula only applies when R^* is at least as larger as r_s. When the photon is emitted at an infinitely large distance, there is no red shift.

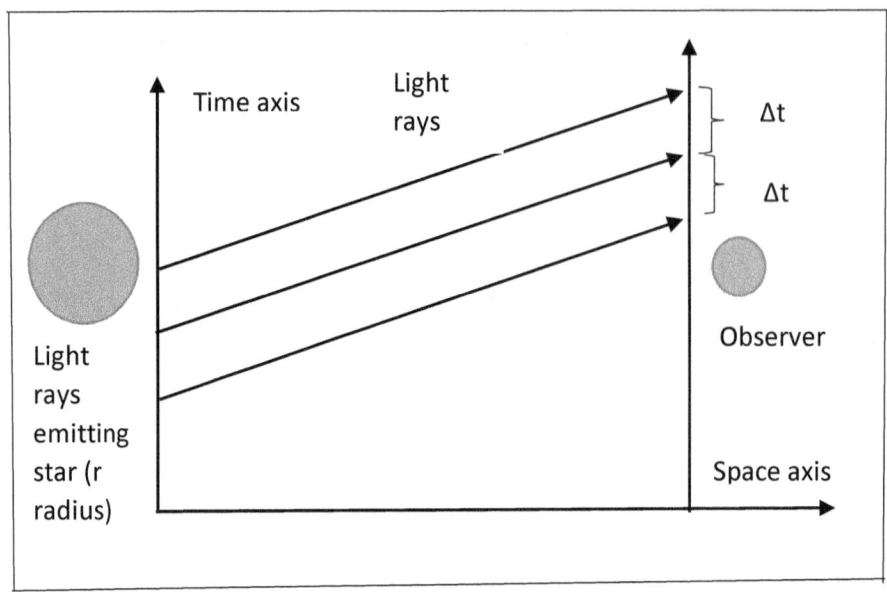

Figure 7.2.4.A: Gravitational red shift (zero for this case) VS radius of the light emitting star

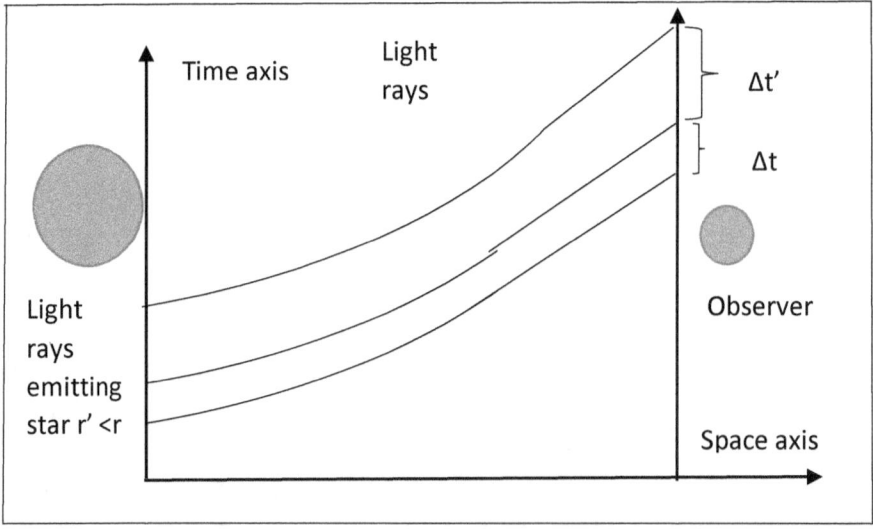

Figure 7.2.4.A: Gravitational red shift (positive) VS radius of the light emitting star

Gravitational red shift in stars

Photons those emitted from a stellar surface on a star of mass M and radius R, have expected to have a red shift equivalent to the difference in gravitational potential. With G the gravitational constant, this potential at the stellar surface is zero at infinity.

The coefficient $G/c^2 = 7.414* 10^{-29}$ cm/g. for the Sun, $M= 2.3 *10^{23}$g and $R = 1.394 * 10^{11}$ cm, so $\Delta\lambda / \lambda = 1.23* 10^{-6}$. In other words, each spectral line should be shifted towards the red end of the spectrum by a little over one millionth of its original wavelength.

In addition, observations of much more massive and compact stars such as white drafts have shown that, Einstein shift occurs and also within the correct order of magnitude. Recently also the gravitational red shift of a neutron star has been measured from spectral lines in the x-ray range. The result gives the quantity M/R, the mass M and radius R of the neutron star. If the mass is obtained by other means, one can measure the radius of a neutron star in this way.

Gravitational red shift (photo origin:
https://en.wikipedia.org/wiki/Gravitational_redshift)

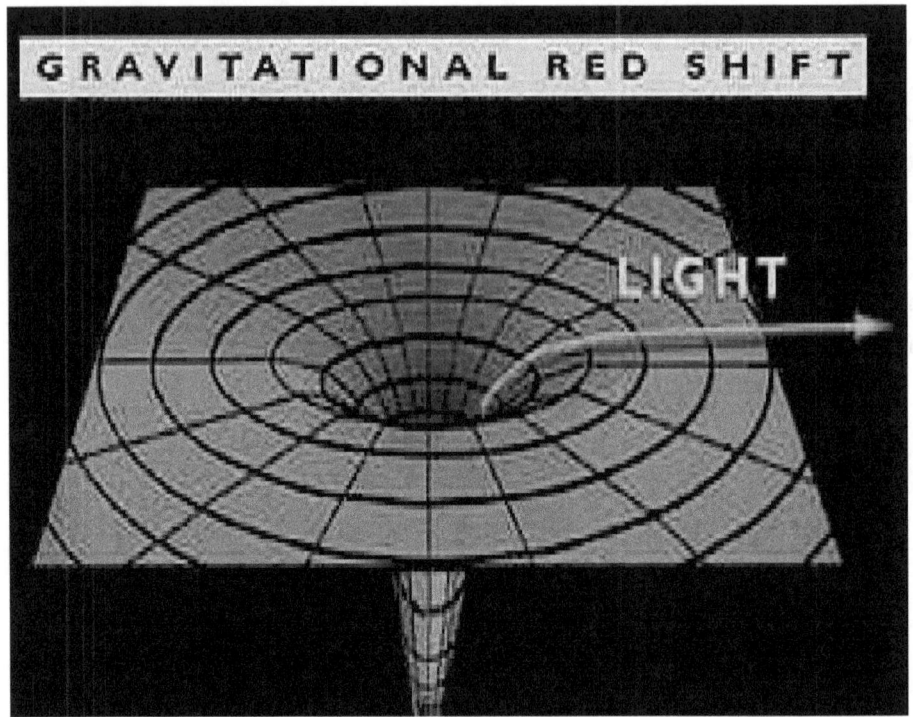

Figure 2.2. (Infinite gravitational red shift)

Photo origin: http://startswithabang.com/

7.3.Methodology

3.1. Infinite gravitational red shift

Schwarzschild equation has related with the Schwarzschild radius of an object. Whenever we want to deal with a star like object such that the radius of that object is varying with space-time (i.e. The radius of the star may get Schwarzschild radius), it is suitable to use Schwarzschild equation.

$$C^2d\tau^2 = (1- (r_s /r))c^2dt^2 — (1- (r_s /r))^{-1} dr^2 –r^2(d\theta^2 + \sin^2\theta d\varphi^2)…..7.3.1$$

7.3.1 is the Schwarzschild equation.

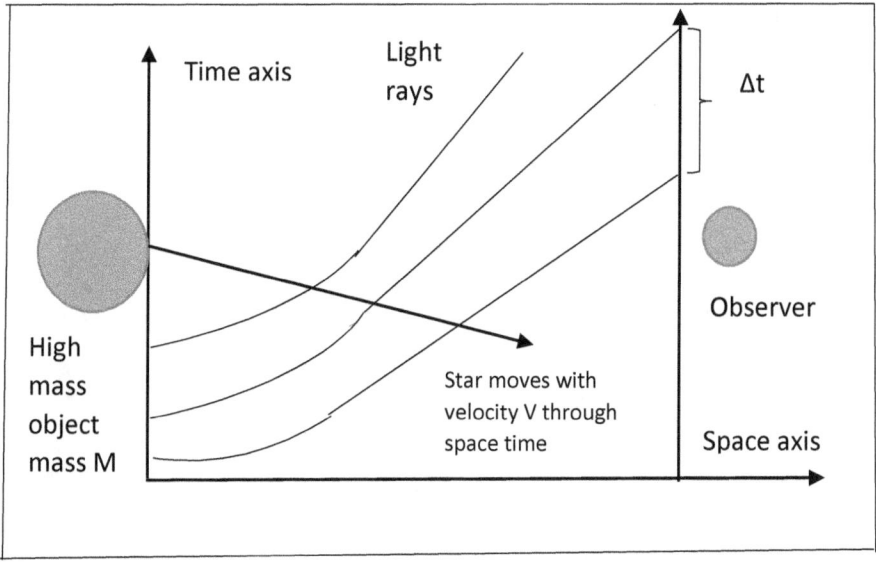

Figure 7.3.1. (Light ray's paths of infinite gravitational red shift and the motion of the massive object)

This figure 7.3.1.shows 'how the star S_1 causes the infinite red shift (curved lines) due to the high mass of star S_1 (Mass M). And also have considered that this star S_1 has reached its critical radius value of a black hole.

When exploring that relation, we want to find: 'what is the equation that indicates the relationship between the velocity of the star and the red shift although the radius of the star attained the critical value of the black hole radius- the light rays going outward from the star will receive by the observer who is at rest relative to the star at some moment). i.e. ignoring infinite gravitational red shift.

Consider the Schwarzschild equation related to the infinite gravitational red shift:

$$c^2 d\tau^2 = (1 - (r_s/r))c^2 dt^2 - (1 - (r_s/r))^{-1} dr^2 - r^2(d\theta^2 + \sin^2\theta d\varphi^2)....7.3.2$$

For the non-spiral motion of the objects near the star (that attains the Schwarzschild radius) the equation 7.3.2. can be modified. Therefore the modified Schwarzschild equation is:

$$c^2 d\tau^2 = (1 - (r_s/r))c^2 dt^2 - (1 - (r_s/r))^{-1} dr^2 \ ....7.3.3$$

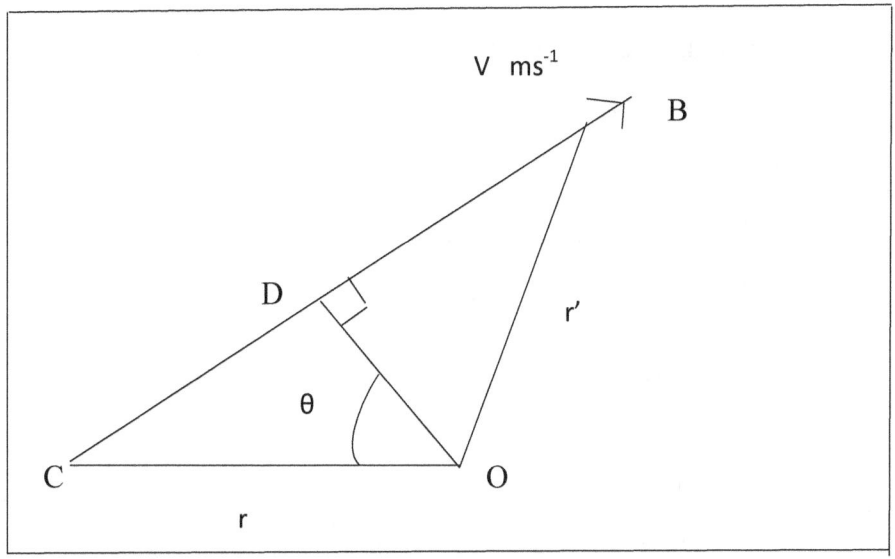

FIGURE 7.3.2: How I find a way to ignore the infinite gravitational red shift

The figure 7.3.2. shows how the light emitting star and the observer have located. The light emitting star is moving with velocity V ms[-1] with respect to the observer who is at O. The observer is rest with respect to the star. The star's initial position is C. And after an arbitrary time, the star is at the position B. The distance between initial position of the star and the observer is r. The distance between the observer and the star's final position (B) is r'. I consider a position that lies on the line CB as D such that the angle COD is θ. Then according to the diagram,

DB = V.dt -r.sin(θ)

$r' = \sqrt{(r^2 .\cos^2\theta + (V.dt)^2 + (r.\sin\theta)^2 — (2.V.dt.r.\sin\theta))}$

$r' = \sqrt{(r^2 + (V.dt)^2 - (2.r.V.\sin\theta.dt))}$

let, $r'/r_s = y$.

Where r_s is the Schwarzschild radius.

$y = \sqrt{(r^2 + (V.dt)^2 - (2.r.V.\sin\theta.dt))} / r_s$

$y = \sqrt{(x^2 + (V.dt)^2 /r_s^2 - (2.x.V.\sin\theta.dt)/r_s)}$

$dy/d\theta = [-(2.x.V.dt/r_s).\cos\theta]*[(1/2).y]$

$dy^2 = [[x.V/r_s.] \sin\theta]^2* d\theta^2*dt^2*(1/y)^2 \ldots\ldots\ldots\ldots\ldots\ldots.7.3.4.$

But, $dy = dr'/r_s$, ; $dr'^2 = dy^2.r_s^2$

By modified Schwarzschild equation (7.3.3.):

$d\tau^2 = (1- (r_s/r')) dt^2 - [(1- (r_s/r'))^{-1}dr'^2/c^2$

$d\tau^2 = (1- (1/y)) dt^2 - [(1- (1/y))^{-1}. r_s^2 .dy^2/c^2]$

$d\tau^2 = (1- (1/y)) dt^2 -$

$\qquad [\{ (1- (1/y))^{-1}. r_s^2 /c^2 \}. [[x.V/r_s.] \sin\theta]^2* d\theta^2*dt^2*(1/y)^2$ (By 7.3.4)

$d\tau^2 = (1- (1/y)) dt^2 - [\{ (1- (1/y))^{-1} \} /c^2 \}. [V.x.\sin\theta/y]^2* d\theta^2*dt^2]$

$d\tau/dt = \sqrt{ \{ (1- (1/y)) - [\{ (1- (1/y))^{-1}\}. [V.x.\sin\theta/(y.c)]^2* d\theta^2] \} }$

When $r = r_s$; $x = 1$.

$d\tau/dt = \sqrt{ \{ (1- (1/y)) - [\{ (1- (1/y))^{-1}\}. [V.\sin\theta/(y.c)]^2* d\theta^2] \} }$...**7.3.5**

Here $y \neq 1$

Therefore although the high massive object and the observer are initially within the Schwarzschild radius, we can calculate the ratio between the proper time (Clock moving with the moving star) and the improper time (A finite value) as measured by the observer for some another position of the light ray emitting star, by using equation 7.3.5.

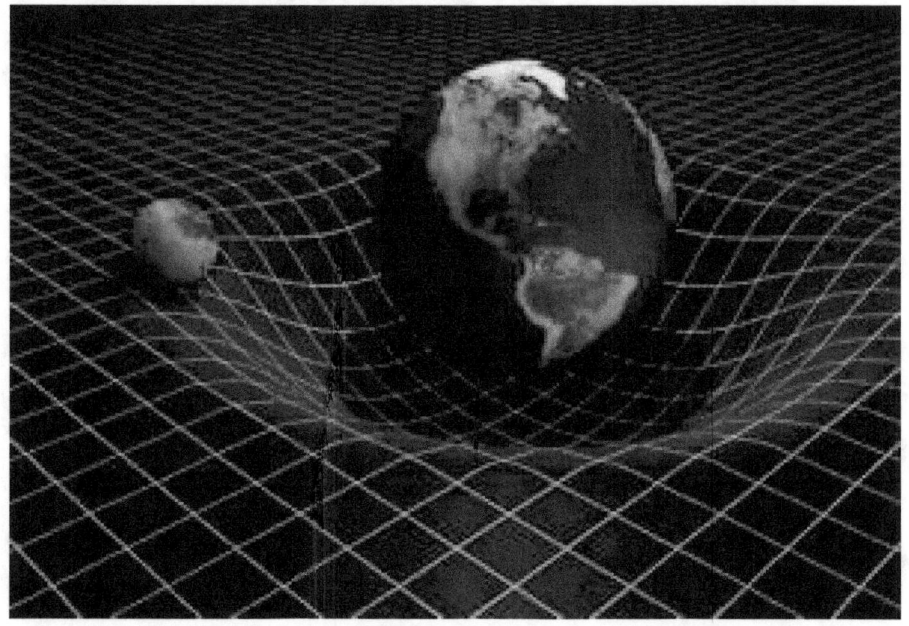

Space time curvature of the Earth like planet (Photo origin:
http://www.waykiwayki.com/2015/07/flat-earth-gravity-is-hoax.html

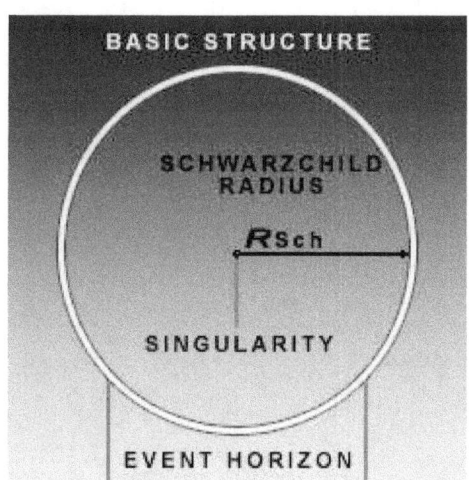

Schwarzschild radius (Photo origin:
http://hubblesite.org/explore_astronomy/black_holes/encyc_mod3_q3.htm
l)

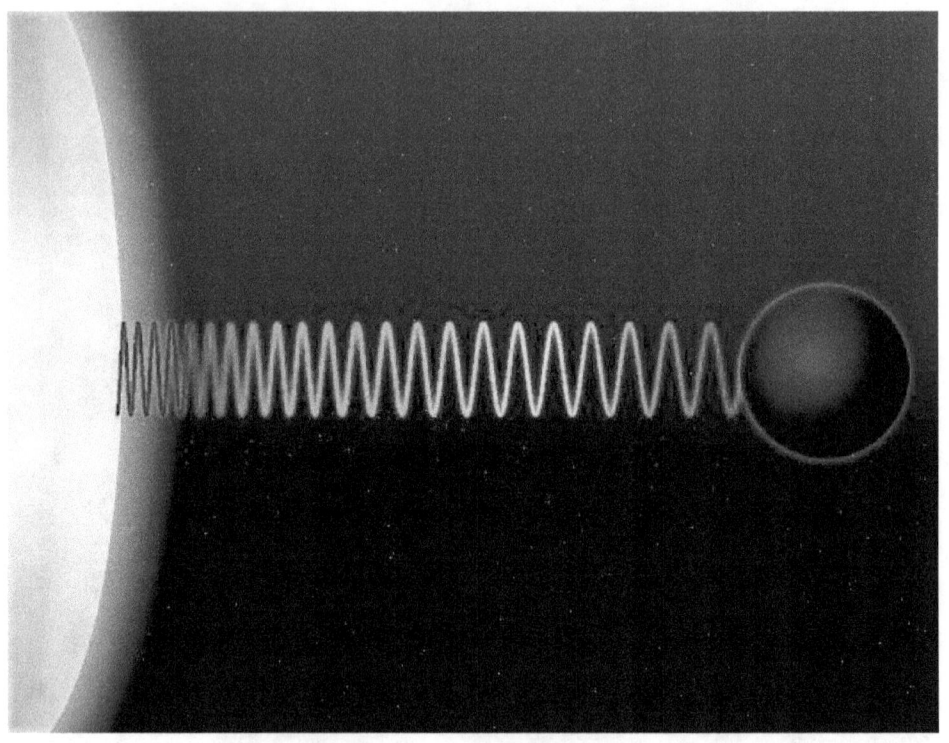

Gravitational red shift effect, photo origin:
https://en.wikipedia.org/wiki/General_relativity

Karl Schwarzschild (Photo origin:
http://www.physicsoftheuniverse.com/scientists_schwarzschild.html)

General theory of relativity (Photo origin: http://shass.mit.edu/news/news-2015-celebrating-einstein-marks-100th-anniversary-general-theory-relativity)

Chapter 08

Wonders of Auroras in the North Pole and South Pole

Our sun is the main heat source for the Earth. There are several contributions from the sun for the Earth. Life on the Earth depends on the sun. But when we consider other planets in our solar system, all other planets do not have any considerable life forms according to present discoveries. But the location of the Earth in the solar system helps to the life of Earth.

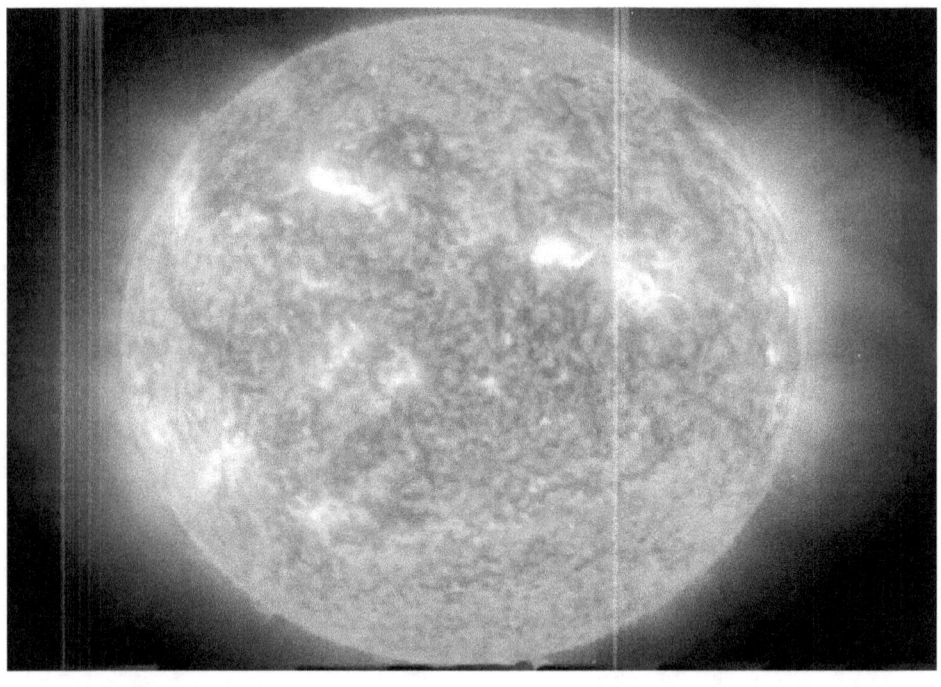

Our Sun (Photo origin: http://globe-views.com/dcim/dreams/sun/sun-05.jpg)

That location is better for life due to the median temperature on the surface. Suitable air pressure is required to keep liquid water on the surface of the Earth. Bur further, there are large number of particles those are coming towards the Earth from the Sun. Due to the magnetic reconnections, solar flares and the explosions on the surface of the Sun, it is allowed the charge particles to release from the surface of the Sun.

Magnetic reconnections of the Sun

Photo credit: M. Aschwanden et al. (LMSAL), TRACE, NASA

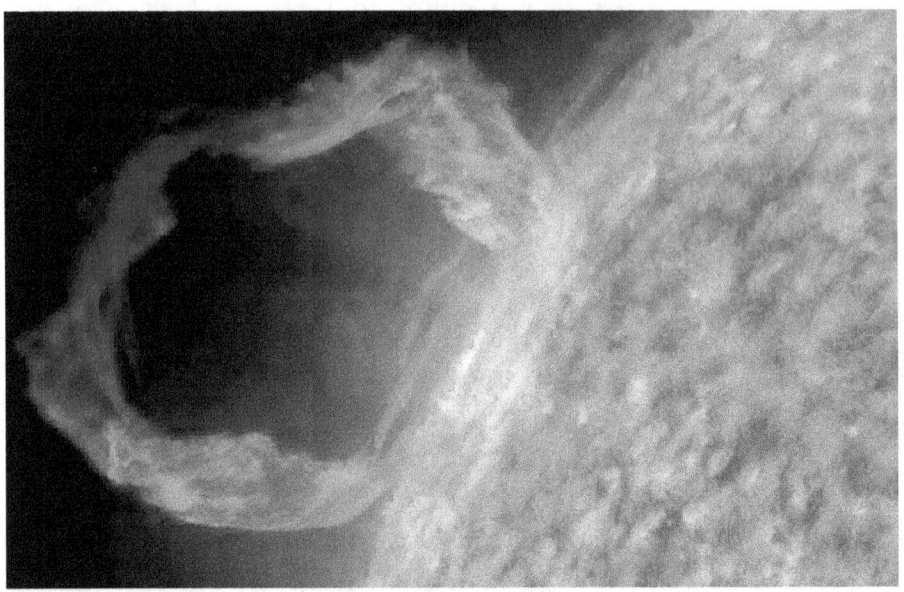

Solar flares of the Sun (Photo origin:
http://www.telegraph.co.uk/news/science/space/9097587/Solar-flares-everything-you-need-to-know.html **)**

There are millions of charge particles those are emitting by the Sun during a second. Those charge particles are highly energized and capable to penetrate the Earth's atmosphere. Also there are large numbers of ultraviolet photons also coming towards the Earth. The ultraviolet photons able to cause cancers and other dangerous effects for the humankind. But the O_3 molecules in the Earth's atmosphere able to cover ultraviolet photons from the Earth's surface. But due to several molecules produced by mankind, it is capable to damage to the O_3 coverings in the Upper atmosphere.

Solar particles towards the Earth (Photo origin:
https://en.wikipedia.org/wiki/Geomagnetic_storm)

But there are several effects those are happening in the Earth's upper atmosphere. As previously mentioned, large numbers of charged particles are coming towards the Earth from the Sun. Usually those charged particles come toward the Earth due to the solar wind. Solar wind contains large number of charged particles those pushed by the energy released by the Sun.

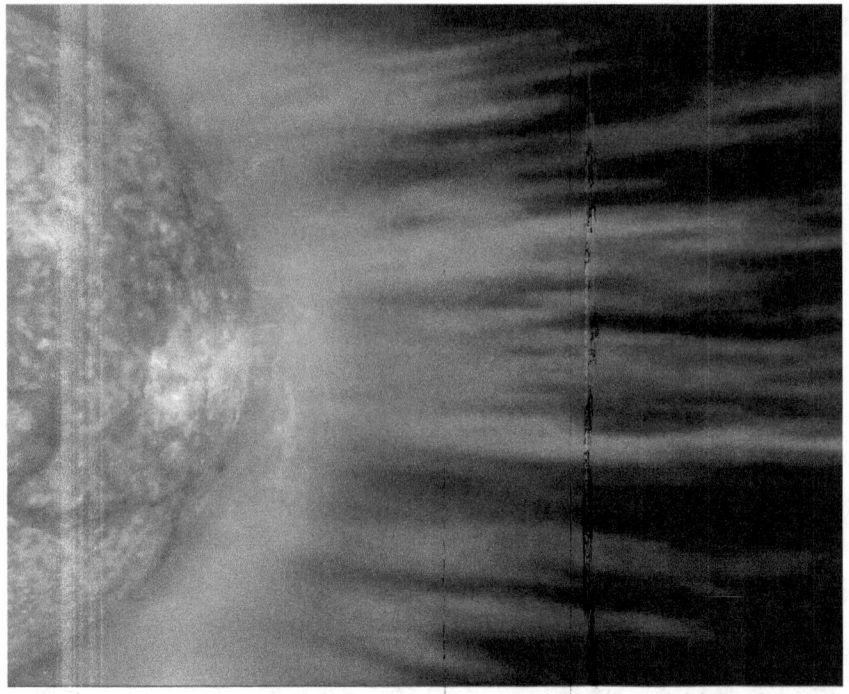

Solar wind (Photo credit: ESA/NASA/SOHO)

The Earth contains powerful magnetic field originating from the North Pole and going towards the Southern pole. The Earth's magnetic fields have the capability to navigate air planes, ships and other land traveling objects. The compasses use the Earth magnetic field to navigate the directions.

Moreover, there is a very interesting effect that happens due to the solar wind particles and the Earth's magnetic field calls as Northern Auroras and Sothern Auroras those are capable to happen in the Southern hemisphere and at the Northern hemisphere. Those are capable to see as Northern Lights and Southern lights. How those auroras are creating?

Auroras near South pole and North pole of the Earth

(Photo origin: http://www.cbc.ca/news/technology/auroras-expected-to-dazzle-again-tonight-1.3125824)

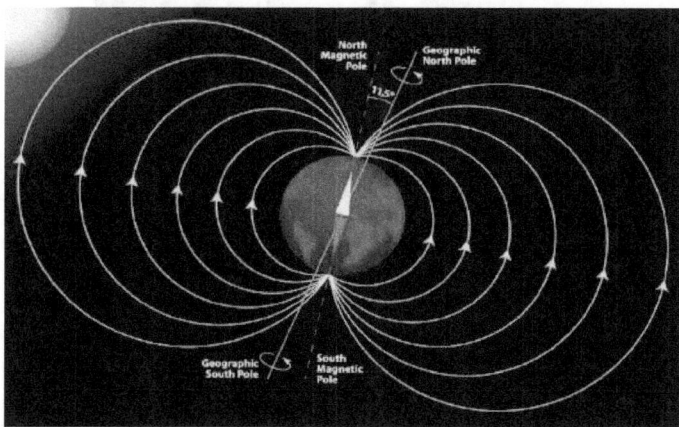

Earth's magnetic field lines (Photo origin:
http://www.crystalinks.com/earthsmagneticfield.html)

The charged particles those are coming towards the Earth from to the solar wind affect by the Earth's magnetic field lines. We all know when a charged particle moves under an external magnetic field, a force acts on the charged particle. The same incident happens under solar wind particles and the Earth's Magnetic field lines.

When the Earth's magnetic field lines act on the charged particles coming from the solar wind, there is a force that is acting on the charged particles. And those solar wind charged particles interact with the Earth's magnetic field lines that produces emitting photons with visible wave length region. Depending on the solar wind particle's energy, the color of the Northern lights and the Southern light vary. But the magnetic field line density is very high near the North Pole and near the South Pole. Therefore there is a very high aurora density near South Pole and near North Pole.

Photo origin: USAF photo by Senior Airman Joshua Strang

Photo credit: Francois Fourie

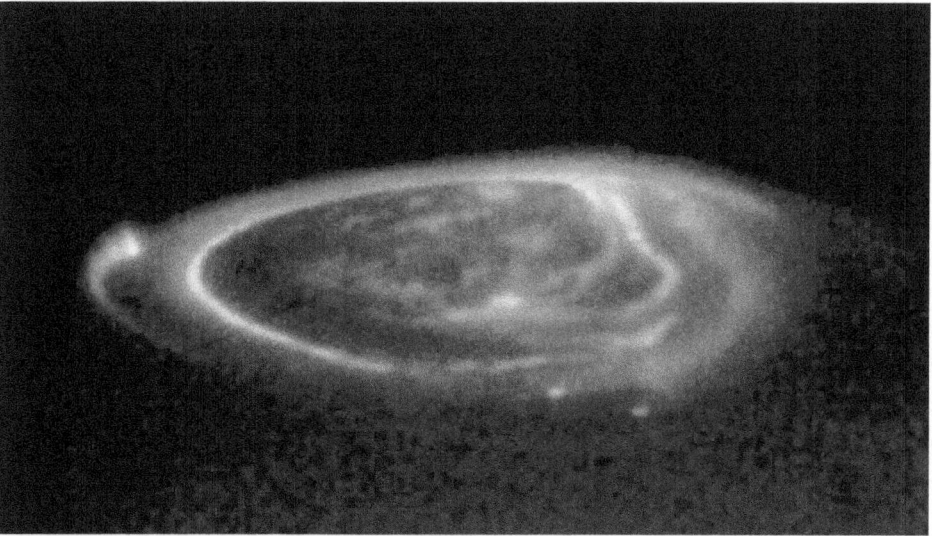

Photo origin: John T. Clarke (U. Michigan), ESA, NASA **(Different colors of Auroras)**

But analyzing the emission spectrum of an aurora, someone able to identify the solar wind molecule elements in each atmosphere area. But if someone analyzes the absorption spectrum of the aurora, he may able to identify the cool gas elements in the Earth's atmosphere near to the auroras directly or out of the aurora.

The composition of solar wind is roughly same content. Therefore a scientist has the capability to analyze the emission and absorption spectrums of auroras in two different regions in northern hemisphere or southern hemisphere. By analyzing the emission spectrum of auroras near two different regions, he may able to identify the photon emitting chemical elements in those two regions separately. The emission spectrum may contain the indications of emitted photons from the solar wind particles and also photon indications due to the upper atmosphere air molecules. But the solar wind particle compositions are roughly same in those two considering regions.

Then the differences of the two spectrums of two different areas of the auroras indicate the Earth's atmosphere air molecule contents in those two regions separately regardless the Solar wind air molecules. Then the scientists can analyze the absorption spectrum. With the usage of that spectrum, he may identify the cool gas elements in the Earth's atmosphere separately in those two regions of auroras. But the unchanged spectrum wavelengths of emission spectrum and the absorption spectrums indicate the chemical elements in the Solar wind particles in each region of Earth's atmosphere.

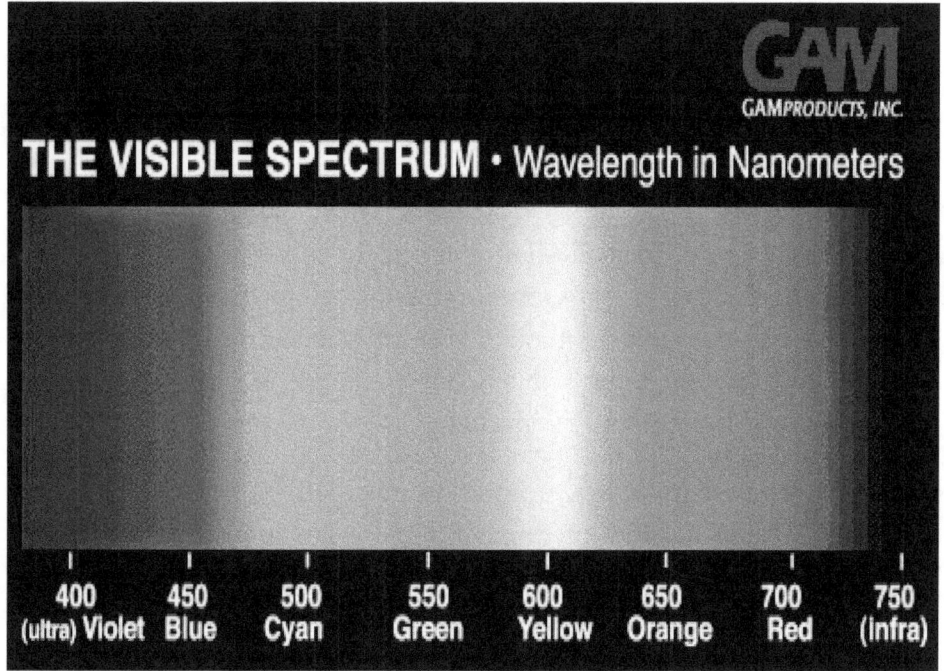

Visible Spectrum

Photo origin:
http://www.gamonline.com/catalog/colortheory/visible.php

Chapter 09

Possibilities of the existence of life on the moon

There are several notions in Science related to the starting of life on the Earth. Some scientists argue that the civilization on the Earth has established by some kinds of Aliens who might came from another planet in the universe. They might came from a planet in our solar system or beyond the solar system (Although present science hasn't revealed the existence of life on a planet in our solar system, there might some life in the solar system before million time ago).

According to present notions, in order to grow life on some planet, there are several chemicals and environments those are needed. As examples, O_2, liquid water, sufficient temperature and etc. But yet, scientists didn't able to prove the existence of all those requirements on some planet in our solar system. But, they try to discover some planet that satisfies those requirements which is in our solar system since the discovery is so useful for the existence of life on Earth.

Future life on moon (Photo Credit: Asa Schultz)

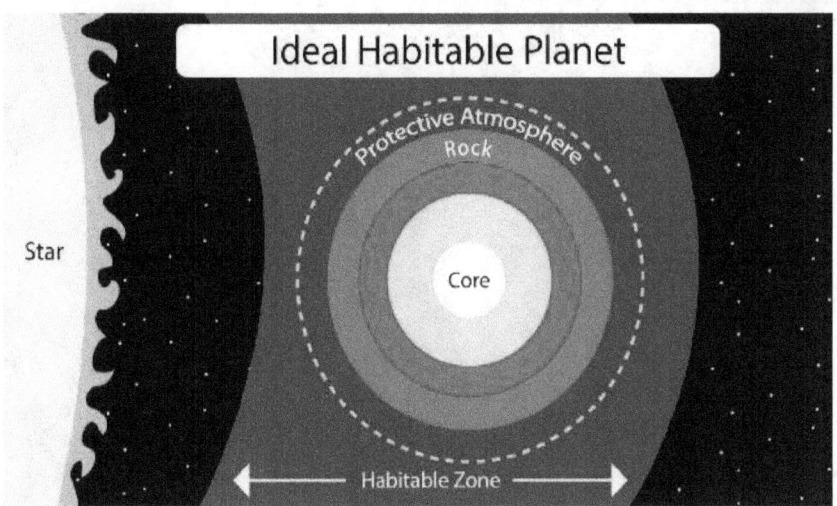

Possibility for the existence of life on a planet (Photo origin:
http://learn.genetics.utah.edu/content/astrobiology/conditions/

We know observers on the Earth can see only one side of the moon. Past scientists focused their major attention to the observable side of the moon. But, there are several researches have done regarding the un-observable side of the moon recently. Several space missions have done in order to carry out the secret features and the chemicals those are existing on the un-observable side of the moon.

Photo origin: www.youtube.com

Liquid H_2O is a major requirement for the existence of life. Therefore, first people have to study about the possibilities of the existence of the liquid water on the moon. For the existence of liquid water, there should be a considerable pressure that is greater than the critical pressure of the liquid water. Moreover, the temperature should obey the above condition. We know that the moon's atmosphere doesn't contain considerable air molecule density. Because, the moon's gravitational field hasn't any capability to catch the escaping air molecules from the moon's atmosphere. Therefore, usually scientists may argue that there won't any considerable atmospheric pressure in order to keep the air pressure that is greater than the critical air pressure for the existence of liquid water on the surface of the moon. But, if the moon's gravitational field is considerably large on some specific region on its surface, then there may be a possibility for the existence of liquid water. But, in order to verify that without any doubt, scientists have to study the absorption and emission spectra of those regions using the highly sensitive scientific equipments.

The chemical compounds on that high gravity area on the surface of the moon, may influence for that unusual physical feature of the surface of the moon. Especially, not only the liquid water, O_2 and sufficient temperature needs for the fine existence of life on moon. We know thorough the solar wind, there are large number of high energy charged particles those are coming towards the Earth from the Sun. Also the Ultra-Violet waves come towards the Earth from the Sun those are harmful for life on Earth. But, the Earth's magnetic field and the Earth's O_3 contents have abled to ignore or reduce the damage due to those rays, for the mankind. But the moon hasn't any strong magnetic field that is capable to turn-off or catch those solar wind particles those are coming towards the moon. Moreover, moon hasn't enough atmosphere air molecule density that is capable to reduce the harmful effects due to Ultra-Violet rays coming from the Sun.

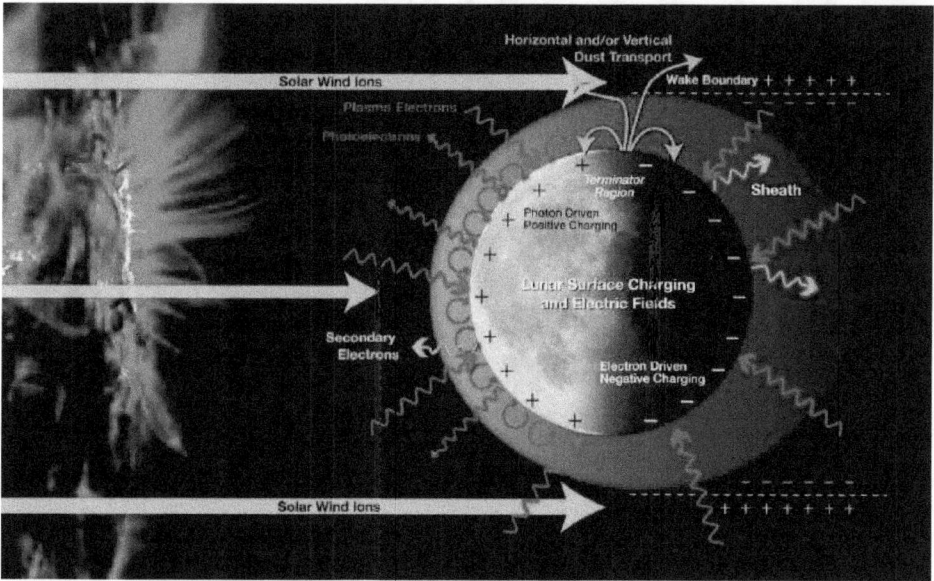

Solar winds towards the moon those are harmful

(Photo Credit: Jasper Halekas and Greg Delory of U.C. Berkeley, and Bill Farrell and Tim Stubbs of the Goddard Space Flight Center)

But, according to my own argument, if scientists able to place some asteroid on an orbit around the moon (That placing asteroid should contain some kind of living bacteria. i.e. there should be a capable environment on the surface of that asteroid for the existence of those life), then in future, the environment of the surface of the moon would help more for any life forms on the surface of the moon. Because the growth of the bacteria would release several useful chemical compounds those are essential for the growth of the life on the moon.

Asteroid around the moon (Photo origin:
https://www.newscientist.com/article/dn23039-nasa-mulls-plan-to-drag-asteroid-into-moons-orbit/)

Chapter 10

Generate electric power using the radiations those are emitted by our living environment

10.1. Concepts introduction

- We know that any object which is in our environment always emits radiation whenever the temperature is greater than 0 K.
- We plan to generate electric power using those radiation emit by the objects which are in our environment.
- And also we have to apply some theorems in Maxwell's electromagnetic theory as, Faraday's law, photoelectric theory and etc.
- Also we have to spend some time for the practical aspect.

  ➢ We know that any object which has the temperature greater than 0 K, emits radiation.
  ➢ But usually the radiation that emit by the objects in our environment ; has very high wave length and has very low frequency.
  ➢ Since the frequencies of those radiations are relatively low (that emit by the objects in our environment), by photoelectric theory, the frequency of those radiations may less than the cutoff frequency for photoelectric emission.
  ➢ Therefore using those radiation those emit by the objects in our environment , we can't carry out electrons from a metal.
  ➢ Now we have to use some procedure in order to amplify the frequency of those radiation as well as have to amplify the amplitude of those radiations.

> ➢ Also we have to use some circuits in order to generate sufficient electric power as well as have to apply some theorems and some concepts in Maxwell's electromagnetic theory , Faraday's law , photoelectric effect and etc.

❖ This new way of generating electric power should able to give us large amount of electric power capacity as well as it should carry a simple procedure whenever the public people are using.

10.2.Concepts:

1. Detect the electromagnetic radiations those emitted by objects which are in our environment.

    a. There are so many electromagnetic radiations in our environment.

    b. We have to catch those EM radiations those are existing within our environment.

    c. We have to use some special technique in order to disturb electric field as well as in order to disturb the magnetic field those are related with EM radiations emitted by objects in our environment.

    d. By disturbing to electric field and magnetic field; we can change the directions of those radiations toward a single area (Just like toward a single detecting component): I use the term 'EM wave detector'.

e. Then we have to amplify the frequency as well as the amplitude of those detected EM radiations.

f. I think we can use superposition principle in order to amplify the amplitude as well as the frequency, by using large number of low frequency and low amplitude EM radiations those are emitted by the objects around us.

g. Although we get small number of superimposed EM waves relative to the number of detected EM waves, those waves have high frequency and high amplitude rather than the detected EM radiations.

h. We have to collide those amplified EM waves with some target metal (like Cu) that has high loosely bounded electrons density.

i. Then we get huge number of electron beams according to the photoelectric effect.

j. Since we amplified the frequency of EM radiations, we get high energized electrons.

k. Since we amplified the amplitude of EM radiations, we get large number of electrons per unit time per unit area.

l. Now we have to bring those beams of electrons to a narrow beam. For that we can use some technique to keep the narrow feature of electron beams.

m. We can use some pointed positively charged distributions in the opposite side of the target metal.

n. We can place some material between the positively charged distributions and the target metal such that electrons should not be able to travel through the material but the coulomb's force should be able to act through the material (To limit the volume that the electrons can travel)

o. Then we can get narrow beams of electrons.

p. We have to apply some voltage in order to carry the electron beams as a current.

q. Now, due to the electron beams, there is some weak magnetic field that should appear.

r. Then by disturbing that induced magnetic field, we can generate some extra electric current by using some metal wires according to Faraday's law.

❖ If we could carry out this experiment, this procedure will contribute to generate electric current as some part of it. Because normally every object within our environment emits EM radiation.

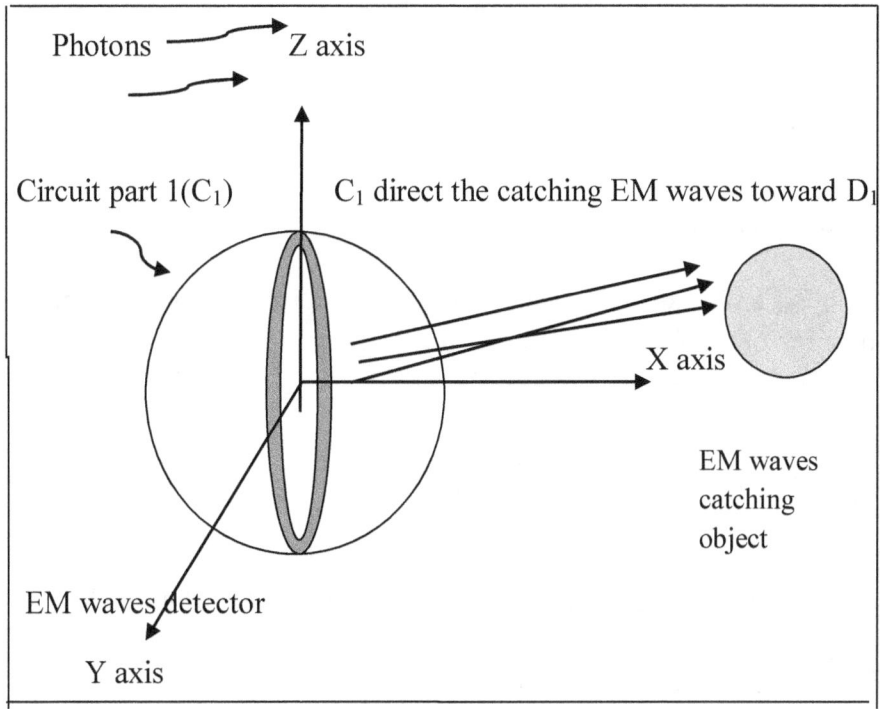

Figure 11.1 (Catching the EM waves)

We use a component as shown above in order to direct the EM waves towards the EM wave detector those are propagating arbitrary; through the space time. This component (C_1) contains some highly charged rings, such that the sphere consists large number of such rings. Such that the center of each ring lies at the center of the sphere. (i.e. the angle 0 to 2π). But between each charged ring there should be some non- conducting material. Then the radiations propagating through space time feel the electric field from each highly charged rings. But each charged ring consist high charge density in every side relative to the origin. But, particular EM wave feels the electric field as most strongly; whenever that feeling electric field is the nearest path to a C_1. Then we know that, a large amount of traveling

EM waves should direct towards the detector as its traveling direction. But, specially there should be some calculated distance between C_1 and D_1. i.e. There should be a definite relative positions between each other.

According to the theorems in electromagnetic fields we know that when the magnetic fields of traveling EM waves direct toward a perpendicular direction, then it makes the traveling direction of each EM wave towards the detector.

We know that the equation for the electric field of an electromagnetic wave is given by

$E = E_m \, \hat{j} \sin(kx - \omega t)$. And this equation implies that the travelling direction of the EM wave is in the x direction.

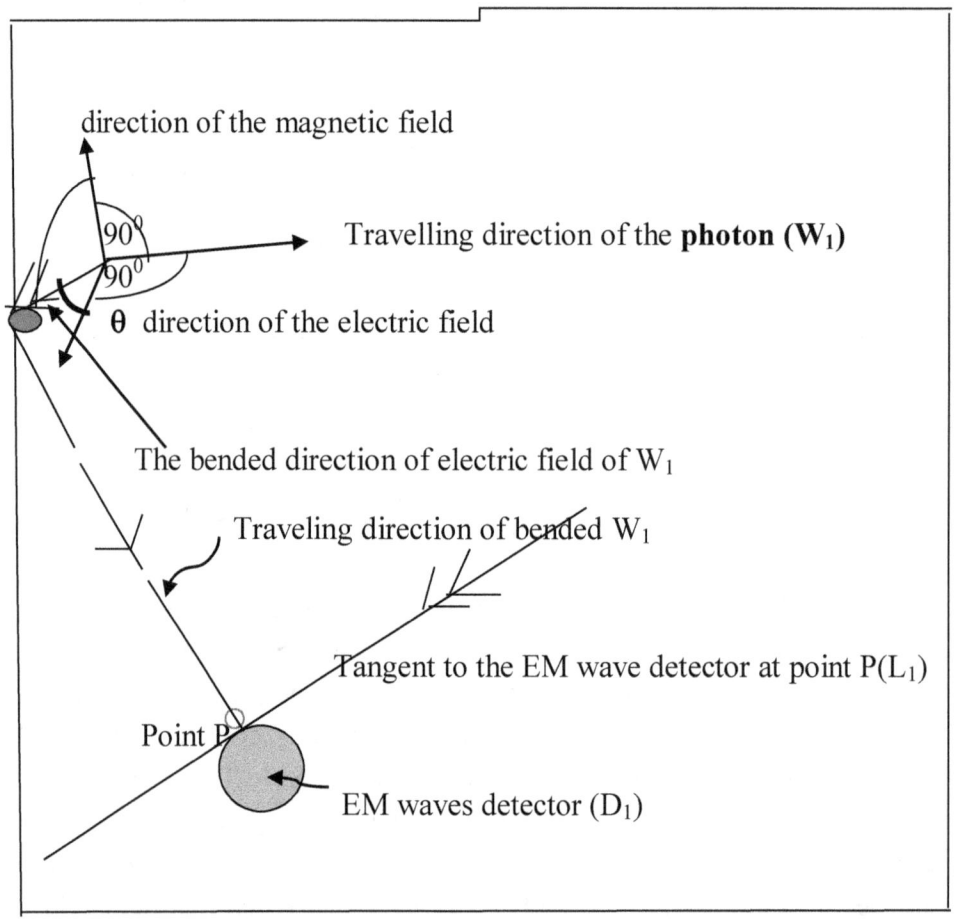

direction of the magnetic field

Travelling direction of the **photon (W_1)**

90^0
90^0

θ direction of the electric field

The bended direction of electric field of W_1

Traveling direction of bended W_1

Tangent to the EM wave detector at point $P(L_1)$

Point P

EM waves detector (D_1)

Figure 11.2

In order to bring the traveling direction of the W_1 is toward the detector, we have to change direction of electric field of W_1 ; such that the electric field of W_1 should parallel to L_1. Thus have to bend the electric field of W_1 according to $E_1 = E^{\wedge} \cos\theta$. Then

$E_1^{\wedge} = E_m \, j^{\wedge} . \sin(kx\cos\theta - \omega t)$. And also the magnetic field of W_1 should convert as

$B_1^{\wedge} = B_m k^{\wedge} . \sin(k. \, x\cos\theta - \omega t)$. Where $B_1^{\wedge}$ is the magnetic field of the bended wave (W_2) of the initial wave W_1. And, $E_1^{\wedge}$ is the electric field of the bended wave (W_2) of the initial wave W_1. Then we know that the traveling direction of the wave W_2 should towards to the EM wave detector.

In order to bend the electric field and the magnetic field of W_1, we have to apply some mechanisms in electromagnetic theory. And those mechanisms should depend on the directions and the magnitudes of the electric fields of traveling EM waves. And also the distance to C_1 from the EM wave (As shown in Figure 11.1). I have mentioned about the theoretical procedure related to this in order to get a rough idea of the experiment.

But, We can keep cycle of C_1 components, i.e. set of C_1 components as a cycle and also we should keep D_1 detectors as two cycles; inside the C_1 cycle as well as outside the C_1 cycle . Then we can detect more number of EM waves. This highlighted part would carry the previous EM wave detecting procedure very simple.

After catching the EM waves by the detectors, we have to amplify the frequency as well as the amplitude of detected EM waves. In order to amplify the frequency and the amplitude of the EM waves, we use the superposition concept. After classifying detected EM waves according to the frequencies and amplitudes ranges, using some modern techniques , we are ensure to apply the superposition concept. Finally we get EM waves such that those EM waves have high energy as well as high intensity. Then those EM waves would allow us to use as incident photons to do the photoelectric effect. By keeping pure Cu plate as the target we would be able to get out large amount of electron beams easily.

This procedure is so easy; because there are so many EM waves. i.e. Those EM waves are so common.

Therefore, I have concluded the related concepts and the procedure of creating the electric current as above and it is a challenge to build up the circuit in a laboratory.

Chapter 11

Travelling to parallel universes

11.1.Introduction

We know, that waves contain matter properties as well as wave properties. Also matters contain wave properties and matter properties as well. When we consider the wave properties of matter; we can consider the De Broglie wave equation for the wave length of the matter wave. The De Broglie wave length equation describes the relationship between the De Broglie wave length of a matter wave and the velocity of the matter wave.

The De Broglie wave length equation is

$\lambda = h / (m.v)$ where λ is the De Broglie wave length of the matter wave, h is the Planck constant and m is the mass of the matter and v is the velocity of the matter wave.

If, we consider Quantum Mechanics principles then we face the fact below. We know a matter wave is capable to appear in different quantum states. If we consider a **plane matter wave** $\Psi(r,t)$ then by quantum mechanics we can express this wave function as

$$\Psi(r,t) = C_1. \Psi_1(r,t) + C_2. \Psi_2(r,t) + \ldots\ldots\ldots + C_n. \Psi_n(r,t)$$

Where $C_1, C_2, \ldots\ldots\ldots C_n$ are complex constants. And $\Psi_n(r,t)$ is the wave function in n^{th} quantum state . $n \in N$.

Then we have the probability of finding the wave particle in n^{th} quantum state $P_n = \Psi_n . \Psi_n^*$ can be written as

$$\Psi_n = k_n . e^{-i.(kr-\omega.t)}$$

Then, $P_n = (k_n . k_n^*) = |k_n|^2$. But if we consider a matter wave with all the possible quantum states that matter wave can appear; then the summation of P_n's equals to 1. That means,

$$\sum_{i=1}^{n} |k_i|^2 = 1$$

11.2. Applications

If my own idea is correct, then this may use to get an idea in order to travel to parallel universes. Final conclusion describes about the largest De Broglie wave length associated with travelling procedure to a parallel universe. If we would be able to find a procedure to travel to parallel universes; then our science and technology would grow up to some highest position. And with this highest position of science and technology, we would be able to explore the universe and universal concepts and theories easily than present. And also the fuel consumption would become very low value for the particular procedure or we would be able to find some extra-ordinary ways to travel from one place to some another space which is far away from the first potion; through a warm hole or through something like that, that we will be able to find in the future.

Sources:

In order to derive those concepts, I used one concept in pure mathematics, I used some few concepts in Quantum Mechanics and I used few concepts in particle physics. The theorems I used to derive the final formula: pure mathematics, quantum mechanics and particle physics those I learnt as a student at the University. And also the things, those I studied at the

University of Colombo. And I didn't use any other fact or concept or theory that other one has used before: which are related with travel to parallel universes.

11.3.Content

Case 1:

We know that if we consider the whole universe ,then for any matter wave; the value should equals to

$$\sum_{i=1}^{n} |k_i|^2 = 1$$

But when we consider the whole universe, then $n \rightarrow$ infinity. Then we get the statement

$$\sum_{i=1}^{\text{infinity}} |k_i|^2 = 1$$

Let assume whole the possible quantum states for the matter wave lie within our universe. i.e. there is no any parallel universes related with the considering matter wave (for the existence).

Lets consider k_i's in the form $|k_i|^2 = (1/2^i)$ where $i \in N$. Where the value $|k_i|^2 = (1/2^i)$ always less than 1 ; for each $i \in N$. But using pure mathematics, we have the relation

$$\sum_{i=1}^{n_0} (1/2^i) \quad \text{does} \quad \text{not} \quad \text{equals} \quad \text{to} \quad 1 \quad \text{exactly.}$$

(We have to consider very sensitive measurement values for the experiment or the theorem)

Then we have a contradiction. By the obtained contradiction, we can conclude the below fact: there exists such a parallel universe which is related with our considering matter wave. (There exists at least one such a parallel universe).

Case2:

Now we consider the case; the all possible quantum states for the existence of the matter wave not lie in the whole universe. That means all the possible this universe's quantum states for the matter wave are in some finite region of space of this universe.

That means we consider 'n' number of dimensions; those are valid according to our usual sense about the finite dimensions.

Then we can assume that there are n number of quantum states within that finite region of space. And, we have the summation for n finite number of quantum states:

$$\sum_{i=1}^{n} (1/2^i) \quad \text{does} \quad \text{not} \quad \text{equals} \quad \text{to} \quad 1 \quad \text{exactly.}$$

Then we have proved the required result for **case 1,** again for this **case 2** also. That means we can conclude that there exists at least one parallel universe, which is associated with the matter wave what we consider. Then we have the result; for any finite region of space (That means we consider 'n' number of dimensions; those are valid to our usual sense about the finite dimension) there exist at least one parallel universe which has associated with our matter wave. And we should aware that we can use particular matter wave such that for all i $\in$ N:

$$| k_i |^2 = (1 / 2^i) \text{ where } i \in N .$$

11.4.Conclutions:

Then we can consider, parallel universes can apply for any place in our space- time (within our universe). That means we can go to a parallel universe, from any place in our usual space-time within our universe. And in order to travel to a parallel universe, there should be some physical rules which have associated with our parallel universe travelling. Then we should consider the fact: what are the rules those apply with travelling to a parallel universe, from any point within our universe '.

We know that when the De Broglie wave length is below from some value, then the associated frequency becomes a very high value. That means the associated energy becomes a very high value. Whenever the associated energy becomes a very high value, then the kinetic energy and associated potential energy also becomes very high value. Then we know that the travelling to a parallel universe is more easy. That means whenever the associated De Broglie wave length is small value, travelling to a parallel universe becomes more easy. i.e. One condition apply for travelling to a parallel universe is: De Broglie wave length should below than some critical value. Now we consider the case: what should be that critical De Broglie wave length that applies for travelling to a parallel universe.

We have the De Broglie relation: $\lambda = h / (m.v)$

where λ is the De Broglie wave length of the matter wave, h is the Planck constant and m is the mass of the matter and v is the velocity of the matter wave.

139

Then consider a matter wave that has used to travel to a parallel universe. Then we know that associated De Broglie wave length:

$\lambda_d = h / (m. v_\varphi)$ Then we have the relation :

$v_\varphi = d(\varphi(r, t)) / dt$.

Let's consider the difference between 1 and $\sum_{i=1}^{n}(\frac{1}{2^i})$ as Δp. Let's take the associated matter wave with the probability Δp is

$\varphi(r, t) = k_p . e^{i.(kr-\omega.t)}$ ....(*). Where k_p is a complex constant. Since the probability of the matter wave $\varphi(r, t)$, lies not in our universe, we can conclude that it has gone to another universe. Therefore the matter wave $\varphi(r, t)$ lies in another universe (That has probability Δp in the other universe)

Then $\lambda_d = h / (m. v_\varphi) \rightarrow \lambda_d = h / (m. k_p. \omega. -i . e^{i(kr-\omega t)})$.

$\rightarrow \lambda_d = h.i / (m. k_p. \omega. e^{i(kr-\omega t)})$.

$\rightarrow \lambda_d = h.i. / (m. k_p. \omega. e^{i(kr-\omega t)})$

$\rightarrow \lambda_d = h.i / (m. \omega. \varphi(r, t))$ 11.1

But , we know that

$d(\varphi^*(r, t)) / dt = k^*_p .i.\omega.e^{-i(kr-\omega t)} = v_\varphi^*$ (By *)

$\rightarrow k_p.e^{i(kr-\omega t)} .v^*_\varphi = i.k_p. k^*_p .\omega$

$\rightarrow \varphi(r,t) = i. \omega. \Delta p / v^*_\varphi$ 11.2

Then by equations 11.1 & 11.2

$\lambda_d = h.i .v_\varphi^* / m.\omega.i.\omega.\Delta p$

$$\lambda_d = h.v_\varphi^*/ m.\Delta p. \, \omega^2 \dots\dots\dots\dots\dots\dots\dots\dots\dots \quad 11.3$$

Because we know,

$$\{ k_p. k_p^* = \mid k_p \mid^2 = \Delta p \}$$

Therefore by equation 11.3, we have

The largest De Broglie wave length associated with travelling to a parallel universe is

$$\boxed{(\lambda_d) = h.v_\varphi^*/ m.\Delta p. \, \omega^2}$$

When the associated De Broglie wave length travel to a parallel universe is becoming smaller; then by the above result

$$(\lambda_d) = (h. V_\varphi^*)./ (m. \Delta p. (2.\pi.f)^2) \rightarrow$$

$$\lambda_d. f = (h. V_\varphi^*) / (m. \Delta p. (4. \pi^2 .f)) \rightarrow$$

$$V_\varphi = (h. V_\varphi^*) / (m. \Delta p. (4. \pi^2 .f))$$

But,

$$\boxed{V_\varphi / V_\varphi^* = h / [m. \Delta p. (4. \pi^2.f)]}$$

Chapter 12

Nuclear Fusion

Big Bang theory explains about the starting and evolution of our universe. According to the big bang theory, at the moment of big bang there was an infinite energy density as well as very high temperature. But, under such unstable physical conditions we know no any atom can exist (specially, under such temperature). But just after big bang happened, universe started to expand rapidly through space time. With the expansion of the universe, the temperature began to decrease rapidly. After some seconds passed, the physical conditions those were in the universe allowed to create the fundamental matter particles. Nature began to create the lightest atom initially (Hydrogen atom is the lightest atom). Because under such relatively high temperature and under such physical conditions Hydrogen is more stable. And, decreasing the temperature went forward and forward. The forward decrease of temperature allowed to create more hydrogen atoms. After the universal temperature decreased up to some value, those Hydrogen atoms were get together and could create an atom that has more mass rather than Hydrogen atom. But the consecutive atom that has more mass rather than Hydrogen is Helium (He). And after that also universal temperature was decreasing. Creating He atoms was proceeded. After some universal time passed, some He atoms could create more massive atoms called Li, Fe, Be and etc. And that should be the starting point of the existence of galaxies, planets, stars, nebulas, quasars and surely human beings also.

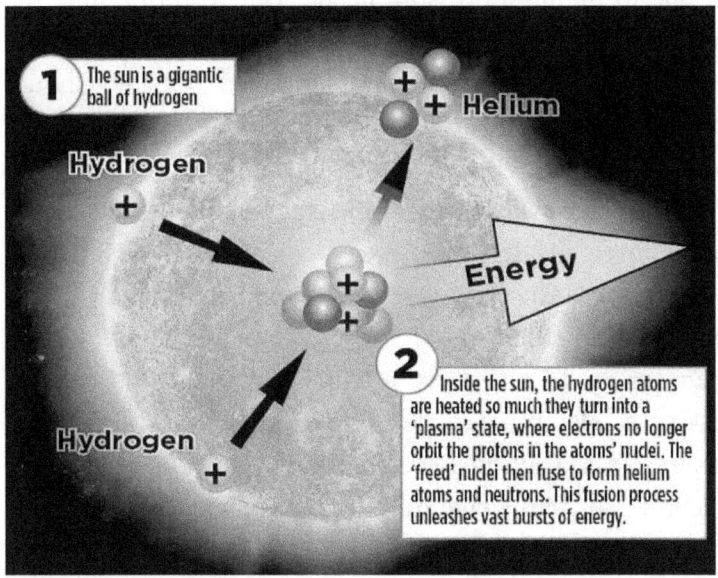

Nuclear fusion inside the Sun (Photo origin:
http://www.dailymail.co.uk/news/article-2587072/Eureka-How-magic-doughnut-fakes-sun-save-planet-But-Chinese-thanks-billions-spend-eco-power-gravy-train.html**)**

We know that inside every star, there is high temperature. And star should be an energy self - generating object. But, the birth of most stars belongs to the old age eras of the universe. And at that time universe filled with high Hydrogen and Helium atomic density. Due to that reason, the major chemical elements of stars are H and He. In particular, we consider our own star 'the Sun'. Sun and usually all other stars are glowing and spread the heat and electromagnetic waves to outer space. But, the energy required should generate by itself. The atoms inside the star (Basically H) use to generate energy with the nuclear fusion. Although our usual living environment can't do nuclear fusion, inside sun, it is possible. Because, Sun has very high pressure and very high temperature inside. And under such conditions nuclear fusion is possible. Three H atoms get together in order to create one He atom which is more massive than a H atom. When that procedure is happening high energy should release to the outside of the star as heat or electromagnetic waves. And there should be a force directed to outside due to the high pressure inside. But with the time, the useful atomic density of H would decrease. Then the above He atoms creating procedure would slow. Then the pressure outside should gradually decrease. But there is an extraordinary force that acts inside the star call gravitational collapsing. Whenever the outside pressure decreases, the temperature inside decreases and ensure the star to collapse under the unstable gravitational force inside. Whenever the star is collapsing, the temperature inside again starts to increase. Then under such high temperature, He atoms get together in order to create more massive atoms like Li atoms. Whenever the Li atoms are producing there should be high energy that is releasing as electromagnetic wave or as heat. Although the number of H atoms are decreasing inside Sun or any other star, they have a procedure (Nuclear fusion) to generate energy again by it- self. But nuclear fusion proceeds under high pressure and high temperature conditions. But the energy power release in nuclear fusion is higher than nuclear fission comparing same mass amount of fuels. But usually nuclear fusion is not practicable within our living environment.

But after billion years has passed, the gravitational force inside would stronger than radiation pull to outside the star. Then some stars become black holes with the time flowing and some others become white drafts and red drafts depending on the mass of the old star.

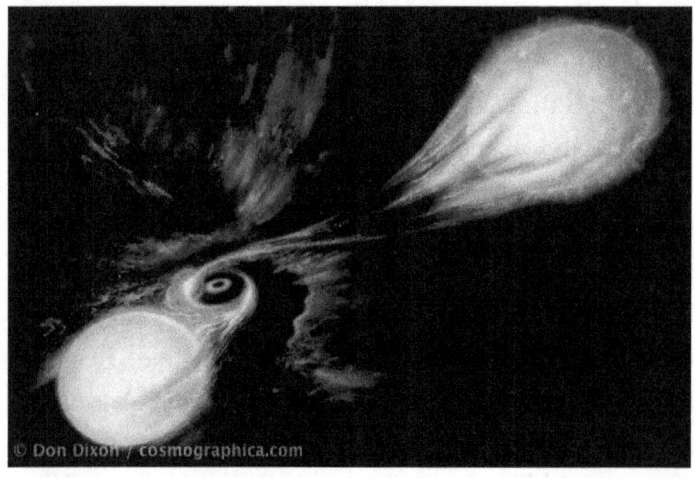

Black hole creation (Photo credit: Don Dixon/cosmographica.com)

Chapter 13

Range for the refraction Index of a material

In this section, I consider a block of some **material with refraction index n.** And, at t=0, a light ray starts to propagate from the position d=0. And at t= t_0 , the light ray comes to the position d=d_0. Also at the moment t=t_0 ,the block of material is moving with velocity U_0 in the direction of light ray propagating. Then at the moment t=t_0, the light ray and the block of material collapse. The situation as below.

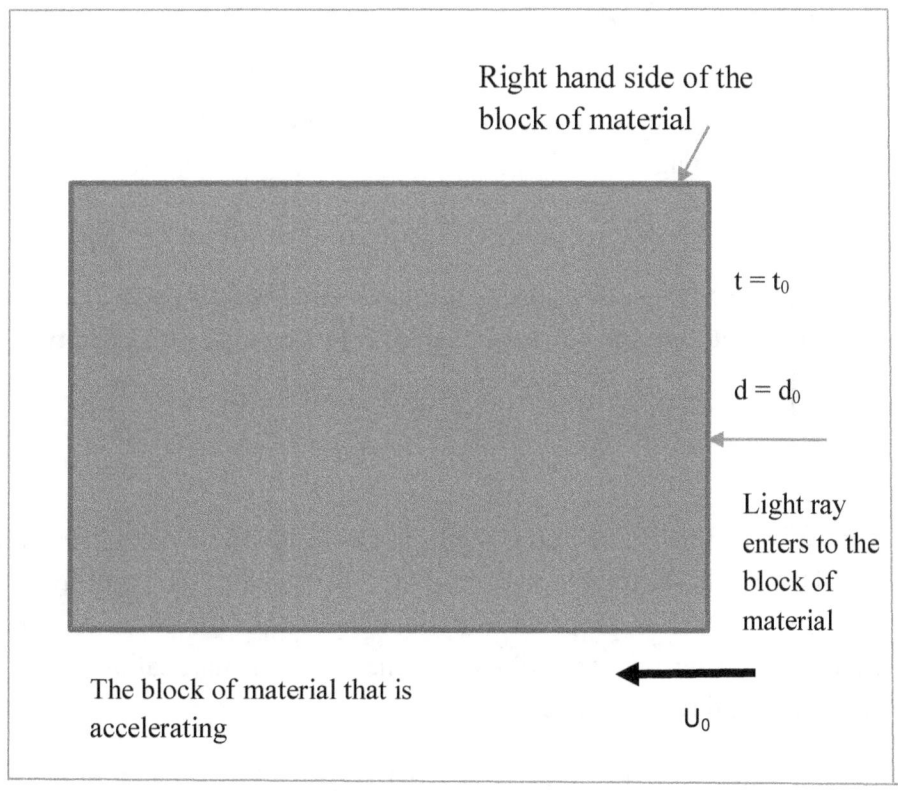

Diagram 13.1 (The situation that the light ray enters into the block of material and block of material is just moving with velocity U_0

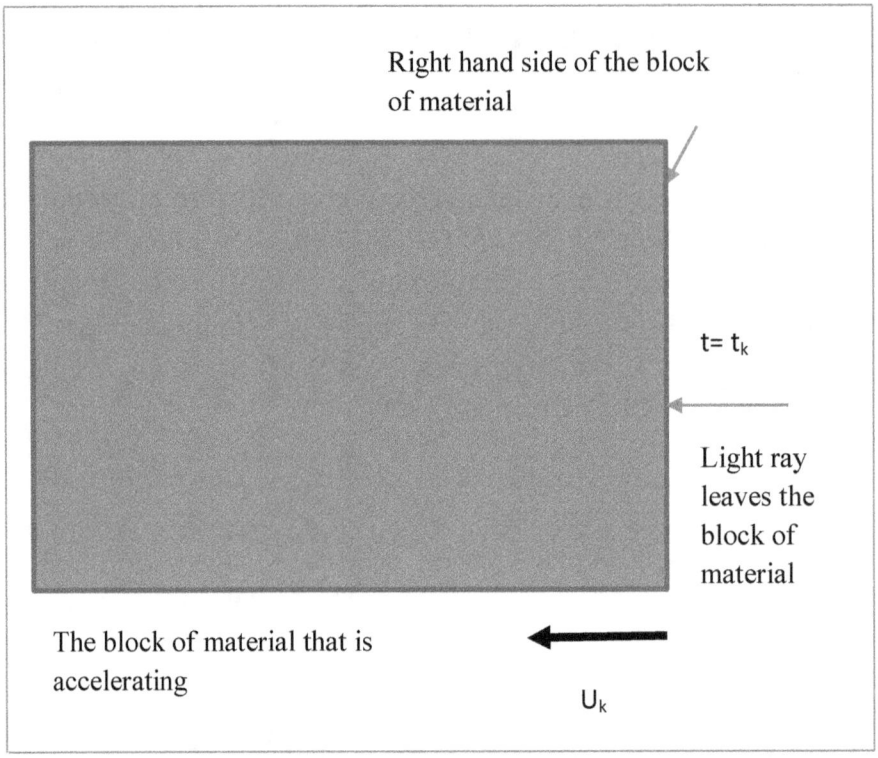

Right hand side of the block of material

t= t_k

Light ray leaves the block of material

The block of material that is accelerating

U_k

Diagram 13.2 (The situation that the light ray leaves the block of material for the first time and block of material is just moving with velocity U_k)

But we know that the speed of the light ray inside the block of material is C/n. That implies after the position d=d$_0$, the speed of the light ray should C/n. But in order to propagate the light ray within the block of material after the moment t=t$_0$, the velocity of the block of material and the value of C / n should related as below.

$$C / n > U_0 \ldots\ldots\ldots 13.1 \ ; \ (C / U_o) > n \geq 1 \ldots\ldots 13.1$$

146

Because the light ray should able to move inside the block of material. In order to move with the block of material, light ray should have a velocity (C/n) greater than U_0

Therefore, there is a range to the value of refraction index of some material as shown in the equation 13.1 (Although the material is upon the surface of the Earth, the material has some moving velocity). And let we consider that the block of material is accelerating.

Consider the case for some U_k ; [$k > 0$; such that each $U_k > U_0$; $k \in R$] $\rightarrow$ the light ray come into the position such that the block of material leaves the light ray (that means block of material passes the light ray) at some moment. **An observer staying outside the frame of incident (at rest: relative to the frame) would measure the length of time such that the light ray propagated inside the block of material (by using the Lorentz's transforms) : $\Delta t'$**

Let dt_i is the time duration that the block of material moved with velocity dv_i. But during the time duration that the block of material has velocity U_0 to velocity U_k , light ray is inside the block of material. Therefore the time spent by the light ray inside the block is same as the time duration spent by the block of material to move with velocity U_0 to U_k. Therefore the time duration of the block of material dt_i as measured by an outside observer = $dt_i / \sqrt{(1 - (V/C)^2)}$. Therefore the total distance traveled by the block of material during dt_i = [$dt_i / \sqrt{(1 - (V_i/C)^2)}$] * dV_i. Here during dt_1 time duration the block of material moves with U_1 velocity. During dt_2 time duration the block of material moves with U_2 velocity. ....And during dt_k time duration the block of material moves with U_k velocity.

Therefore,

$$\left[\int_{U0}^{Uk} \{ \left[\int_{i=0}^{i=k} dti \right] / \sqrt{[1 - \left(\frac{v}{c}\right)^2]} \} dv \right] / [Uk - U0] \quad = \Delta t'$$

$$\Delta t' = \left[(\Delta t / (U_k - U_0)) \right] * \int_{U0}^{Uk} 1 / \sqrt{1 - \left(\frac{v}{c}\right)^2} \, dv$$

Where, Δt is the time difference that is according to the block of material frame itself measured from $t = t_0$ to $t = t_k$ (That means the proper time of the block of material).

$$\Delta t' = [C . (\Delta t) / (U_k - U_0)] . [\ Sin^{-1} (U_k / C) - \ Sin^{-1}(U_0 / C)]$$
.............................13.2

The distance travelled by the light ray within the block of material with refraction index n:

$$\Delta d = (C / n) . \Delta t \dots\dots\dots\dots\dots\dots13.3$$

Because the time duration measured by the block of material is same as the time duration spent by the light ray ($= t_k - t_0$). Because the velocity of a light ray is independent from the observer. Therefore the time taken to move the same distance is independent from the observer. Therefore in equation 13.3, we can apply Δt.

Let $\Delta d'$ = the total distance that the right hand side of the block of material travelled (= the distance travelled by any point of the block of material- because the block of material is a rigid body)

$$[\ C \ . \ (\Delta \ t)]. \ [\quad Sin^{-1} \ (U_k \ /C) \quad - \quad Sin^{-1}(U_0 \ / \ C) \quad] \ = \ \Delta d' \ldots\ldots\ldots\ldots\ldots\ldots 13.4$$

But by the equation 13.3 $\rightarrow$ $(C/n)^* \Delta t = \Delta d$. But we know that $n > 1$ (Because the block of material does not contain the air). Therefore

$(\Delta d). n > \Delta d$. Therefore , $C (\Delta t) > \Delta d$ 13.5

But by using the equation 13.4 we see that , since the value of (C/n) is usually very large value; and the light ray leaves the block of material at the velocity U_k (i.e. $U_k > C/n$). That implies the value of U_k is so much larger than the value of U_0 .(because $(C / n > U_0)$). Therefore we can conclude that , the term $(sin^{-1}(U_k /C) - sin^{-1}(U_0 /C)) > 1$. Therefore, the value of $(\Delta d')$ is greater than the value of $(C .\Delta t)$.

Therefore, $\Delta d' > (C . \Delta t)$.................................13.6

Therefore by the equations 13.5 and 13.6 we can conclude that the value of $\Delta d' > \Delta d$...13.7

By using the equation 13.7 , eventually we see $\rightarrow$

Although the distances (Δd) and ($\Delta d'$) measured by the same observer, the values of those two distances are not same. (But, normally we can decide the values of those two distances are same). Because the distance that the light ray travelled is same as the distance that the block of material's right hand side travelled. Therefore, we get a contradiction.

Therefore, we can conclude that there should be some paradox with our used basic and fundamental physical quantities.

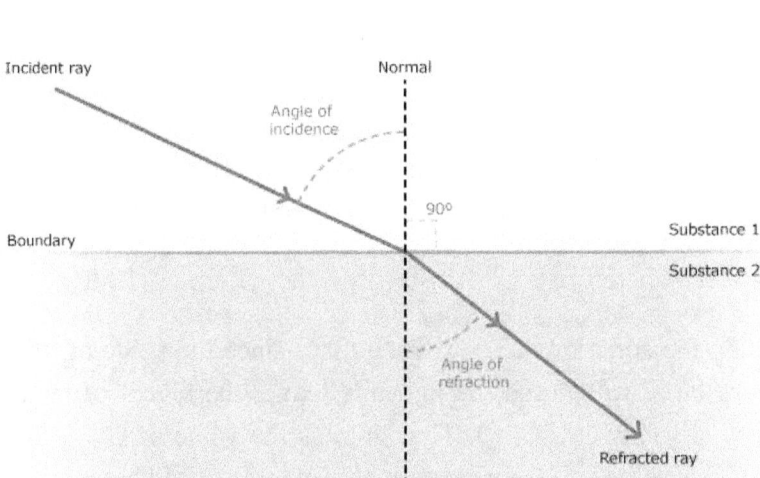

Refraction of light rays depending on the refraction index

Photo origin: http://sciencelearn.org.nz/Contexts/Light-and-Sight/Science-Ideas-and-Concepts/Refraction-of-light

Appendix

- When a person 'A' moves with velocity V with respect to another person B, there is a time dilation in each person's reference frame. The time dilation equation of person A, can be written as below.

$$t = t' / \sqrt{1 - (V/C)^2}$$

Where t is the improper time for the considering person A, and t' is the proper time for the considering person A. Also, C is the speed of light.

In relativity, proper time is time measured by a single clock between events that occur at the same place as the clock. And improper time is time measured by a single clock between events that occur at the different place as the clock.

- Relativistic Doppler formula:

$$f_s / f_0 = \sqrt{[1 + (V/C)] / [1 - (V/C)]}$$

Where, f_s = frequency of the wave the source emitted

Where f_0 = frequency of the wave the observer received

V = relative velocity between the observer and source of waves

C = speed of light

- In mathematics, **a metric** or distance function is a function that defines a distance between each pair of elements of a set. A set with a metric is called a metric space.

A metric on a set X is a function (called the distance function or simply distance)

$$d : X \times X \rightarrow [0,\infty),$$

Where $[0,\infty)$ is the set of **non-negative** real numbers (because distance can't be negative so we can't use **R**), and for all x, y, z in X, the following conditions are satisfied:

1. $d(x, y) \geq 0$
2. $d(x,y) = 0 \Longleftrightarrow x = y$
3. $d(x, y) = d(y, x)$
4. $d(x, z) \leq d(x, y) + d(y, z)$

- **Tensors** are geometric objects that describe linear relations between geometric vectors, scalars, and other tensors. Elementary examples of such relations include the dot product, the cross product, and linear maps. Euclidean vectors, often used in physics and engineering applications, and scalars themselves are also tensors.
  ** **Please refer the definitions of tensors in detail**

References

- http://ffden2.phys.uaf.edu/212_fall2003.web.dir/Eddie_Trochim/assumptions.html
- http://lefteriskaliambos.wikia.com/wiki/EINSTEIN%E2%80%99S_WRONG ASSUMPTIONS_IN_SPECIAL_RELATIVITY
- http://voyager.egglescliffe.org.uk/physics/relativity/post1.html

- http://www.upscale.utoronto.ca/PVB/Harrison/SpecRel/SpecRel.html

- https://en.wikipedia.org/wiki/Relativistic_Doppler_effect

- http://hyperphysics.phy-astr.gsu.edu/hbase/relativ/reldop3.html

- https://en.wikipedia.org/wiki/Energy_level

- http://hyperphysics.phy-astr.gsu.edu/hbase/quantum/qnenergy.html

- https://en.wikipedia.org/wiki/Lorentz_transformation

- http://www.sparknotes.com/physics/specialrelativity/kinematics/section3.rhtml

- http://arxiv.org/abs/0808.0126

- http://journals.aps.org/prd/abstract/10.1103/PhysRevD.8.1633

- Solar wind Wikipedia, Van Allen Radiation Belt Wikipedia, http://www.epa.gov/ozone/science/sc_fact.html

Author's Photo (K.H.K. Geerasee Wijesuriya)

www.ingramcontent.com/pod-product-compliance
Lightning Source LLC
Chambersburg PA
CBHW080618190526
45169CB00009B/3224